ON MARINE CULTURE

A Cultural Glimpse into One of the
World's Finest Military Organizations

Thomas Smith

Chapbook Press

Schuler Books
2660 28th Street SE
Grand Rapids, MI 49512
(616) 942-7330
www.schulerbooks.com

On Marine Culture: A Cultural Glimpse into One of the World's
Finest Military Organizations

ISBN 13: 9781943359974

Library of Congress Control Number: 2018935447

Printed in the United States by Chapbook Press.

To my mother, *in memoriam,* and my father.

To Julie and our beautiful daughters, Peyton and Kylie.

And to Marines across the globe who have constructed the unrivaled military record and reputation of the Marine Corps. Their professionalism, selflessness, and *esprit de corps* have made the Marine Corps not only a revered institution but a state of mind.

There are only two kinds of people that understand Marines: Marines and the enemy. Everyone else has a second-hand opinion.

General William Thornson (USA)

Marines I see as two breeds, Rottweilers or Dobermans, because Marines come in two varieties, big and mean, or skinny and mean. They're aggressive on the attack and tenacious on defense. They've got really short hair and they always go for the throat.

RAdm. "Jay" R. Stark, USN;
10 November 1995

Other people tell you what they do. Marines tell you what they are.

Anonymous

Old breed? New breed? There's not a damn bit of difference so long as it's the Marine breed.

Lieutenant General Lewis B. "Chesty" Puller

I have just returned from visiting the Marines at the front, and there is not a finer fighting organization in the world.

General of the Army, Douglas MacArthur
September 21, 1950.

Why in hell can't the Army do it if the Marines can? They are the same kind of men; why can't they be like Marines?

Gen. John J. Pershing, US Army;
12 February 1918

The deadliest weapon in the world is a Marine and his rifle.

Gen. John J. Pershing, US Army

I am convinced that there is no smarter, handier, or more adaptable body of troops in the world.

Prime Minister of Britain,
Sir Winston Churchill

Marines are about the most peculiar breed of human beings I have ever witnessed. They treat their service as if it was some kind of cult, plastering their emblem on almost everything they own, making themselves up to look like insane fanatics with haircuts to ungentlemanly lengths, worshipping their Commandant almost as if he was a god, and making weird animal noises like a band of savages. They'll fight like rabid dogs at the drop of a hat just for the sake of a little action, and are the cockiest SOBs I have ever known. Most have the foulest mouths and drink well beyond man's normal limits, but their high spirits and sense of brotherhood set them apart and, generally speaking, of the United States Marines I've come in contact with, are the most professional soldiers and the finest men I have had the pleasure to meet.

Anonymous

The Marines I have seen around the world have the cleanest bodies, the filthiest minds, the highest morale, and the lowest morals of any group of animals I have ever seen. Thank God for the United States Marine Corps!

Eleanor Roosevelt

Table of Contents

Commander's Intent

U.S. Marines are taught a concept called "commander's intent." It refers to the commander's intent or purpose underlying an order issued to junior Marines. A commander's intent usually comes into play on vague, ambiguous, or incomplete orders. In such events, Marines turn to the commander's intent, which is expressly spelled out by the commander, to help them interpret the order and figure out how to properly execute it to achieve the commander's intent. Well, as the author of this book, I am also its commander. And my intent in writing this book is to acquaint[1] you with Marine Corps' culture. You will learn about Marine culture from a former insider. I was a U.S. Marine officer on active duty in 1988 and from 1991-1995 as a 2nd Lieutenant, a 1st Lieutenant, and finally a Captain. I went through Officer Candidate School (OCS) in 1988, The Basic School (TBS) in 1991-1992, and Naval Justice School (NJS) in 1992. I was stationed at Camp Pendleton, CA from 1992-1995.

I write about Marine culture for three principal reasons. First, I want to recognize what goes largely unrecognized: that Marine culture plays an extraordinary role in causing—or at least mightily contributing to—the extraordinary success of the Marine Corps, one of the great

---

1 By "acquaint" I mean to introduce you to Marine culture. I do not intend to provide an exhaustive or comprehensive study of Marine culture. And even if I did, it still would be no substitute for the real thing—going through Marine training, experiencing its culture, and earning the title "Marine." As Marines like to say, the title "Marine" is only "earned, never given." Reading this book will never do it for you.

institutions of the world (military and non-military). Marine culture also contributes mightily to the extraordinary success of individual Marines, both inside and outside the Corps. In my view, Marine culture, more than any one thing, more than training, more than weapons, more than physical fitness, is what makes the Marine Corps the Marine Corps—and a Marine a Marine. As Thomas Ricks wrote in his important book, *Making the Corps*: "Culture…is all the Marines have. It is what holds them together." I could not agree more.

Second, I believe Marine culture deeply, and often times permanently, affects the behavior and conduct of individual Marines—both inside and outside the Corps. As James A. Warren writes in *American Spartans*, "On the individual level, the transformation [from civilian to Marine] is both profound and, in the vast majority of cases, permanent. Few indeed are the former Marines who do not retain a strong emotional connection to the [Marine Corps]." Tom Clancy, in his book, *Marine*, wrote this about the permanency of the Marine Corps' experience: "Whether [Marines] are in for just a few years, or make it a lifelong commitment…the Corps changes them all for life." If an institution and culture have this kind of profound and permanent effect on its members, it should be identified, explored, and discussed. The fact is, there are important lessons to learn, practice, and adopt from the Marine Corps, regardless of being in the military, in the business world, or in another profession.

Third and finally, I am aware of no other book or monograph exclusively devoted to explicating Marine culture.

Rules of Engagement

Marines are given Rules of Engagement before being sent into harm's way. Accordingly, I think it's only appropriate that the reader, before reading about Marine culture, be given Rules of Engagement. The Rules of Engagement for this book are as follows:

1. You must understand what this book is about—and what it's not about. This book is about my personal observations and opinions of Marine culture. It is not a memoir of my experiences as a Marine officer, although at times I will share some of my experiences in italics to illustrate points about Marine culture. It is not a book about the experience of becoming a Marine, enlisted or officer. There are numerous books in print on that issue.[2] Nor is this book a scientific study of Marine culture. No scientific tests or opinion polls were administered to Marines, former and current. Nor is there any historical research data comparing the culture of the Marine Corps with other cultures, military or non-military. Rather this book simply describes my walk through the Marine Corps, what I saw, what I felt, and what affected me and others. It's a de Tocqueville-like tour, although much more simplistic, through Marine culture.

---

2 See *Making the Corps*, by Thomas Ricks; *Boot*, by Daniel Da Cruz; *Rows of Corn*, by Herb Moore; and *One Bullet Away*, by Nathaniel Fick.

2. You must understand that this book is my view of the elephant. I cannot, and do not, profess to speak for all Marines about Marine culture and the making of Marines and being a Marine. This book is simply my observations and opinions of Marine culture. I analogize it to the proverb of the blind men feeling the elephant. Each one has his own way of feeling, processing, and coming to conclusions about the beast. The same holds true with the Marine experience. While the Marine experience and Marine culture is factual, external, and one seemingly huge thing, it is not the same huge thing to everyone. It is many things at once, depending on one's vantage point. So I'm sure some current and former Marines will quarrel or take umbrage with my interpretation and description of some aspects of Marine culture. My response? They may be right.

3. You must understand that this book, unless specifically referring to a female, exclusively uses the male singular pronoun. This is for a few reasons. First, the male pronoun—he, him, his—was originally and traditionally intended to encompass both a man and a woman. It should be so construed here. Second, the repeated use of *both* male and female pronouns—he or she, his or hers—grows tiring to the reader. Third, statistical preponderance is in favor of the use of the male pronoun: the overwhelming majority of Marines in the Marine Corps are men (approximately 93%). For all those reasons, the male singular pronoun reigns supreme in this book.

4. You must understand that what you see may not be what you get. I do not warrant that everything you read in this book is what you will experience in today's Marine Corps. Marine Corps' culture is ever-evolving. Because of this, no book can be entirely accurate in its descriptions. This book, moreover, is based on my active-duty experiences in the Marine Corps from a long time ago, in 1988, and from 1991-1995. Nonetheless, I am confident that so much of the culture then remains the same today. If this wasn't true, the Marine Corps would not have been able to consistently churn out its high-quality and standardized products— U.S. Marines—since 1775. This indeed is a testament to the core parts of the Marine Corps, including its culture and traditions, remaining the same.

5. You must understand that to become a Marine requires considerable effort, diligence, and persistence. Enduring Marine training is an intense experience, to say the least. The Marine Corps—the Big Green Machine—is a scary monster to many people, particularly to the uninitiated. It is not a democracy doling out rights to its members. Rather it's an autocracy, making people leave the comfort of their homes to report to screaming Drill Instructors, Platoon Sergeants, and Sergeant Instructors, who are all too ready to make them suffer physically, mentally, and emotionally. It's all about owning its members, from head to toe, immersing them in military training 24/7, with little sleep, and testing them with written examinations and gargantuan physical exercises—all while trying to make them learn the rudiments of war, get along, build friendships, and graduate. At its core, the USMC is about preparing people for the horrifying business of battle, i.e., killing human beings, something anathema to civilized people. The USMC is also a meritocracy—a remorseless and brutal one—where you pass the tests, and therefore pass muster, or you get recycled or kicked-out of the Marine Corps. Therefore, getting through Marine training requires considerable endurance, a great will, much energy, and an unwavering focus on the objective: becoming a Marine and being awarded the Eagle, Globe, and Anchor (EGA). In short, you have to *want it*. Badly. What comes to mind here is a quote by President Calvin Coolidge on "persistence":

Nothing in this world can take the place of persistence. Talent will not; nothing is more common than unsuccessful men with talent. Genius will not; unrewarded genius is almost a proverb. Education will not; the world is full of educated derelicts. Persistence and determination alone are omnipotent.

So I tip my hat to those who have gone through the Recruit Depots at Parris Island, South Carolina and San Diego, California, and also to those who have gone through Officer Candidate School at Quantico, Virginia. They have experienced the pain, the loneliness, and the unmitigated fear, regardless of graduating or not. Those proud folks have taken the risk, entered the arena, and looked into the eyes of a lion—a much more commendable circumstance than refusing to leave the comfort zone of one's couch.

On Marine Culture

1

"It's the Culture, Stupid!"

The notion of "culture" gets short shrift in the Marine Corps. Perhaps this is because many Marines think "culture" is too esoteric, too touchy-feely to be treated seriously in the rough and tumble world of the United States Marine Corps. You'll notice this if you talk to a Marine about culture. His eyes will glaze over and his mind will numb. You'll hear something like, "What the hell are you talking about?" Whatever the reasons, these facts remain true: many if not most Marines don't understand Marine culture, and they don't recognize the importance, the omnipresence, and the extraordinary contribution culture makes to the Marine Corps' success and to their own success. If Marines are asked to explain such success, both at the Marine Corps' level and individually, most will say something like this: "It's because of our training"…and…"I was just doing my job." But the word "training" in this context doesn't quite capture "culture," nor is it intended to. Rather, most Marines intend "training" to mean instruction in military tactics, in the use of weapons and equipment, and in physical training. And while this definition of "training" cannot be ignored, for it plays a significant role in the success of the Marine Corps, it is not the entire answer. We know this because many of the world's military forces are similarly trained and equipped but are not considered "elite" military forces, like the Marine Corps. Consider the US Army, British Army, French Army, and German Army. The answer, therefore, is something more. This book argues that the answer can be analogized to an old political aphorism used by James Carville, former President Bill Clinton's presidential campaign advisor: "It's the economy, stupid!" Well, in reference to the extraordinary success of the Marine Corps, one can easily say, "It's the culture, stupid!" In short, Marine culture is the mechanism that makes Marines special—and it creates an atmosphere, an environment, where the whole becomes bigger than the sum of the parts. Tom Clancy, one of our favorite non-military military experts, wrote this about U.S. Marines in *Marine*, one of his nonfiction books:

> Marines are more than the sum of their equipment. They are something special. They take the pieces that are given to them, arrange them in unique and innovative ways...and throw in their own distinctive magic. There is more to military units than hardware. There is the character of the unit's personnel: their strengths, experience, and knowledge, their ability to get along and work together amid the horrors of the battlefield.

So what is this thing called "culture"? Let's begin with formal definitions. *Webster's Ninth New Collegiate Dictionary* defines "culture" as "the customary beliefs, social forms, and material traits of a racial, religious, or social group." In *Making the Corps*, Thomas Ricks defined it as "the values and assumptions that shape its members." In my view, on a less formal level, "Marine culture" means a high-pressure cauldron, where the customs and traditions of the Marine Corps, including values like hard work, selflessness, honesty, integrity, loyalty, courage, discipline, responsibility, respect, courtesy, pride, punctuality, and neatness are infused into its members, from the admission process on—hour after hour, day after day, week after week, month after month, year after year.

Some commentators on the Marine Corps have referred to a Marine's "ethos." This, too, is as an important part of being a Marine. As Tom Clancy writes in *Marine*, "There is an almost undefinable quality [to Marines]. That quality is the Marine Corps' secret weapon. Their edge. That quality is their *ethos*." [Emphasis in original.] George F. Will, in one of his columns on Marine Corps officer training, said this: "[W]hat makes the Corps not only useful but fascinating is…its conscious cultivation of an ethos conducive to producing hard people in a soft age." Consequently, we should understand the concept of "ethos" and its interplay with "culture." Again, let's begin with the formal definitions. *Webster's* defines "ethos" as "the distinguishing character, sentiment, moral nature, or guiding beliefs of a person, group, or institution." Tom Clancy, in *Marine*, defines it as "the disposition, character, or attitude of a particular group of people that sets it apart from others"—or, "in short, a trademark set of values that guides that group toward its goals." In my view, and again on a less formal level, "ethos" means everything that makes up or comprises the end product of a Marine. And the difference between "ethos" and "culture" can be explained by a cause-and-effect example: culture produces ethos. In other words, "Marine culture" is the environment, the method, the device—the cauldron—used to instill, inculcate, and sustain the things that comprise a Marine's "ethos"—i.e., the final characteristics of a Marine.

Of course cultures are not new. They've been around for centuries, and they are a ubiquitous presence around the world: nations, tribes, religions, businesses, schools, jails, and even families have them. In fact, every institution or entity of humans has, at least superficially, a culture—regardless of whether its members choose to recognize it or not. And like fingerprints, no two cultures are alike. Cultures vary in their content, in their precise nature, and how they affect their members. What is more, they are ineffable: they defy precise description, perhaps because they are amorphous, translucent, and, at least to outsiders, complicated. This is all true of Marine culture. I mean, how do you accurately describe Marine culture, or any culture for that matter, especially to an outsider? It's about as easy as getting your arms around a 500-pound marshmallow. For instance, is it good enough to say Marine culture is intense, powerful, controlling, and cutthroat? Well, if so, wouldn't that also describe Microsoft's and Google's culture, including many other businesses and institutions? Now, taking it a step further, we could also say something more descriptive, like the following:

> Marine culture is a place where the whole is bigger than the sum of its parts. It's a place where the individual is subordinated, in the most drastic way, to a higher cause or purpose: accomplishing the Marine mission. It's a place where teamwork, loyalty, hardwork,

and selflessness reign king. It's a place where *Semper Fi* trumps *Semper I*.

This may be fancy and cute and accurate, but is it enough? Hardly. Fancy adjectives and cute sayings won't cut it. And neither will analogies. Thomas E. Ricks, in *Making the Corps*, analogized Marine culture to Japanese culture:

> The culture that the Marines most resemble, oddly enough, is that of Japan. The Marines are almost a Japanese version of America— frugal, relatively harmonious, extremely hierarchical, and almost always placing the group over the individual. "In normal situations, Japanese in principle accept the needs of the group as much as possible," writes the commentator Ryushi Iwata, "while on the other hand, they die and repress their own needs."
>
> Both the Marines and Japanese society operate in a kind of physical and even psychological isolation from the larger world. And central to both cultures is the sense of being locked in a struggle for existence. This hovering threat hones the warrior culture in both the Marines and Japan.

Marine culture has also been analogized to a cult-like religious order. Says James A. Warren in his book, *American Spartans*, a battle history of the U.S. Marine Corps from Iwo Jima to Iraq:

The Corps has been referred to derisively as a kind of military cult, and more dispassionately as a kind of military "denomination." Indeed, there are some striking similarities between the Marines and a religious order. Both require a transformation for full membership, a kind of rebirth. Both require the willing acceptance of a core set of beliefs. Both require an enduring commitment to a cause greater than oneself. For the Marines, stories of Iwo Jima, the Chosin Reservoir, Hue, and Chesty Puller are sacred. They serve as scriptures do for religious groups.

Although these Japanese and religious analogies are helpful in explaining Marine culture, they still do not cut it. To capture the essence of Marine culture, one has to penetrate the culture on a deeper level, a more profound level, and apply critical powers of observation and description. One cannot capture it by adjectives, cute sayings, analogies, or in a paragraph or two.

So what follows is my attempt to capture and describe Marine culture and to bring clarity to ambiguity. To do this, I have subdivided what I believe are Marine culture's salient characteristics into stand-alone chapters. Some of these chapters can be construed as values and virtues comprising a Marine's "ethos," like selflessness, integrity, loyalty, attention to detail, physical toughness, and aggressiveness, and how Marine culture inculcates them into Marines. But other chapters cannot be so construed. Instead these other chapters might contain general descriptions of, say, Marine Corps' history, the Marine Corps' ability to unfailingly crank out high-quality products, the Marine stereotype, and the way Marines speak, et cetera. Although these traits may not be specific values or virtues, they do, in the end, either contribute to making Marine values and virtues or are necessary, in my view, to a proper understanding of the Marine Corps and Marine culture. Finally, it is important to know that the inculcation of a Marine's ethos in Marine culture occurs both formally and informally—in the classroom, in the field, the barracks, the chow hall, the chapel, in trucks, tanks, aircraft, cars, bars, and even over the phone. In short, Marine culture is formless and limitless, extending to every place where Marines congregate, converse, and interact.

2

Elite Force of Winners

The United States Marine Corps, at approximately 184,000 strong, is acclaimed as the world's largest elite military force—larger even than most nations' armed forces. It is a combined-arms, self-contained, versatile force of warriors, equipped with its own tanks, armored personnel carriers, amphibious assault vehicles, jets, and helicopters (attack and transport). The Marine Corps deploys these assets with a finely-tuned symphonic precision, killing its enemies both efficiently and expeditiously.

Since its founding at Tun Tavern in Philadelphia on November 10, 1775, the Marine Corps has become America's "first to fight" 9-1-1 force, equally at home "in the air, on land, and sea," as the Marines' Hymn so accurately points out. Perhaps more importantly, the Marine Corps has become a revered and feared military force, garnering the respect of historians, national leaders, citizens, and, most of all, its enemies. Historian John Keegan has called the Marine Corps "one of the most formidable fighting organizations in the world." Former U.S. Representative and Ambassador Clare Boothe Luce, in Lt. Gen. Victor H. Krulak's book, *First to Fight,* had this to say about the Marine Corps:

> Since the birth of our nation, the steady performance of the Marine Corps in fighting America's battles has made it the very symbol of military excellence. The Corps has come

to be recognized worldwide as an elite force of fighting men, renowned for their physical endurance, for their high level of obedience, and for the fierce pride they take, as individuals, in the capacity for self-discipline.

As evidence of their special place in the minds of US political leaders, Marines are detailed to US Embassies around the world (and also to the President's band, called The President's Own). These highly public assignments contribute to the perception that Marines are special. But what truly makes Marines special is their elite reputation as a revered and feared fighting force of men. This reputation extends far and wide. As Tom Clancy writes in *Marine*: "The Marines have a global reputation. Whether it's fear or respect—probably a little of both—people around the world know exactly who the U.S. Marines are." He also wrote that the most powerful weapon against a potential enemy of Marines is "the fear of what might happen if one had to face a force of American Leathernecks in battle. You see, Marines are mystical. They have *magic*." [Emphasis in original.]

Hugh McManners, in his book, *Ultimate Special Forces*, believes the Marine reputation works to its advantage:

> [T]he fearsome reputation of the United States Marine Corps [is such] that its involvement can have a powerful impact on the enemy, psychologically as well as militarily. Perhaps this is why the Marine Corps still spearheads US forces into battle, and is often given the toughest missions against the most determined enemy.

On Marine Culture

The Marine reputation, of course, is formed in blood and grounded in an illustrious battle history. From its inception, the Marines Corps has taken on the best and worst of the world's military (and irregular) forces—and they have not been found wanting. They have closed with and destroyed enemy combatants from the world's best military units, such as the vaunted German Army at Belleau Wood in World War I; the Japanese Imperial Army at Guadalcanal, Pelileu, Saipan, Iwo Jima, and Okinawa in World War II; the Chinese and North Korean Army at Chosin Reservoir (and the North Korean Army at Inchon) in the Korean War; and the North Vietnamese Army and Viet Cong at Khe Sanh and Hue in the Vietnam War. They have also closed with and destroyed, and generally made easy work of, the world's second-rate military (and irregular) units, such as those from Cuba, Haiti, Dominican Republic, Puerto Rico, Guam, Samoa, Mexico, Honduras, Panama, Nicaragua, Grenada, Iraq (read: Fallujah and An Nasariya), and the Phillipines, to name only some.

Put simply, the Marine Corps has a penchant for winning. And if there is a notion Americans understand and embrace, it is this one. Al Davis, the late, great owner of the LA Raiders football team, famously told his teams: "Just win, baby!" Well, the Marine Corps ascribes to Al's pithy aphorism, for they win. At times, of course, they win ugly, with grievous casualties and countless killed, but they win. Because of their illustrious battle record, and feared reputation, Americans believe that Marines will not only distinguish themselves each time they are deployed against any foe, foreign or domestic, in any clime and place, but that they will also, more importantly, prevail. Such statements as "Tell it to the Marines!"…"Send in the Marines!"…"The Marines have landed and the situation is well in hand!" have become immortalized in the American lexicon.

3

Production of Excellence

The real and untold story of the Marine Corps is not about its elite status, its ability to project force across the globe, its matchless battle record, or even its fearsome reputation. Instead, the real and untold story of the Marine Corps is about its amazing ability, since November 10, 1775, to consistently produce a high-quality and tangible product—a U.S. Marine. This ability, and product, has remained remarkably unchanged over the years. From its inception to today, the Marine Corps has produced Marines who are exceptionally fit, physically and mentally tough, competent in the rudiments of war, professional, and, with some well-publicized exceptions, even virtuous. This consistency is amazing given the pool from which people are taken and made into Marines. In the view of many, American society has, over the years, lapsed morally and turned out increasing numbers of undisciplined, disrespectful, selfish, immoral, indolent, and generally non-cooperative people—people who are more interested in individual rights and money than in having duties and helping others and their country. But the Marine Corps squeezes out these traits and replaces them with new traits and values. Corps values. Marine values. The end product—a U.S. Marine—is almost unrecognizable from its former civilian self. Sure, the exterior has changed, and sometimes significantly, with muscle replacing fat, and a soft, rounded jaw-line giving way to a hard cornered one. But the crucial and life-long changes are less visible and reside on the inside: values and traits like discipline, pride, selflessness, working hard, courage, honesty, courtesy, punctuality, patriotism, and professionalism. Despite being provided flawed material, the Marine Corps continues to transform people. Even more importantly, they continue to win battles and inspire confidence.

The Marine Corps is held in such high esteem that some commentators equate it to Special Forces. In *Imperial Grunts*, author Robert Kaplan, after spending time with Green Berets in Afghanistan and Marines in Iraq, had this to say about the similarities and differences between Green Berets and Marines:

> The Army Green Berets with their beards, ball caps, and Afghan dress were individualist; the Marines, with their extreme, "high and tight" crew cuts and digital camouflage uniforms, were standard-issue company men. And yet they both shared something vital, something which deeply attracted me: the history and tradition of Special Forces and the Marines were in counterinsurgency and unconventional war. Special Forces and the Marines, each in its own way, epitomized military transformation. Both these branches of the military combined nineteenth-century techniques with twenty-first-century technology. Because the lessons of conventional industrial age warfare of the twentieth century did not apply to the War on Terrorism, real military transformation

would come about only when the Big Army and the Big Navy became more like Special Forces and the Marines, rather than the other way around.

4

Insiders and Outsiders

To Marines, who prefer order and structure and black-and-white simplicity, there are two types of people: Marines and everyone else. Insiders and Outsiders. To Marines, you're either a Marine or you're not. You have crossed the Rubicon, paid the high price of admission, endured the ultimate hazing process of shared depravation and ridicule, entered the Mysterious Fraternity of Men Born of Smoke and Danger of Death, and earned the title U.S. Marine—or you have not. If you *have* earned the title, you are granted certain inalienable rights: bragging rights, favoritism rights, the right to discriminate against Outsiders, and the right to get close and bond with Insiders, including even the right to criticize Insiders and the Marine Corps. If you *have not* earned the title, you have none of these rights. You are an Outsider physically, mentally, morally, and spiritually. You are pressing your nose against the glass, wondering what it's like on the inside to be one of The Few. The Proud. The Marines. To Marines, outsiders are non-members and therefore non-entities. Outsiders include Soldiers (especially Soldiers!), Sailors, Airmen, Coasties (members of the Coast Guard), and of course civilians.

On the inside, the Marine Corps is loved, revered and respected—and hated, reviled, and resented. Often at the same time. And often by the same Marine. It's a peculiar paradox to be a Marine: a Marine is likely to say he hates the Marine Corps about as much as he says he loves it. To be sure,

On Marine Culture

Marines love some things about the Corps: the people, the comraderie, the elite reputation, the firing of weapons, the travel—but they hate other things, like the separation from family, the continual 10-, 15-, 20-, and 25-mile conditioning hikes, the cleaning of weapons, the unforgiving nature of the Marine Corps, and the times when they're cold, wet, and hungry, with no warmth, dryness, or food in sight. All this produces the peculiar paradox: Marines love to hate the USMC, a.k.a. the Big Green Machine. Anthony Swofford, in his memoir *Jarhead*, wrote: "Like most good and great Marines, I hated the Corps." Truth be told, the "hate" word is used more frequently than the "love" word. This "hate" aspect of Marine culture is not limited to the Marine Corps, however. It is present in other male-dominated, macho institutions. As one example, author Robert Timberg in his book *The Nightingale's Song,* a book detailing the lives of five famous Naval Academy graduates, described in the book's first section how most Midshipmen hated the Naval Academy and how frequently they used the "hate" word. He titled this first section, "Book One: IHTFP," meaning, as he later writes, "I Hate This Fucking Place." Make no mistake, Marines say the same thing, with the same frequency, but only with changing locations. A Marine's hate moves in lock-step from Quantico and San Diego and Parris Island to Camp Pendleton to Camp Lejeune to Okinawa to Anyplace the Fucking Marine Corps Sends Him. But again, this hate is for Insiders, not Outsiders. Outsiders better watch out if they criticize the Corps in mixed company. One quick way to stop Insiders from criticizing their Corps is to have an Outsider criticize the Corps in their presence. You can rest assured that the Insider will be too busy dealing with the

Outsider's criticisms to proffer any of his own.

One example of a hybrid situation running counter to the Insider-and-Outsider outlook is the Navy Corpsman, otherwise referred to as "medic" or "doc" in the Marines and Navy. The Marines use Navy Corpsmen to treat wounded Marines. Marines do not produce their own medics or doctors. They use the Navy's trained personnel for that mission. And be rest assured, even though a Corpsman is not a Marine, Marines still embrace and respect a person who places himself in harm's way and treats wounded Marines like Corpsmen continually do. Some Marines don't like the concept of having someone attached to their unit who did not go through Marine boot camp, but that initial dislike is usually, in the end, suppressed by the examples of selflessness and courage.

5

You Can't Subdivide a Marine—
At Least Not in Theory

As someone once said, "Other people tell you what they do. Marines tell you what they are." In the Marine Corps, the title "Marine" (and the awarding of the Eagle, Globe, and Anchor device) is the be-all and end-all of membership. And unlike other services, the title applies to both officers and enlisted alike. (The US Army, for instance, confers the title, "Soldier" on its enlisted members and "Army officer" on its officers; the US Navy confers "Sailor" and "Naval officer," the US Air Force "Airmen" and "Air Force officer.") Even more, the title "Marine" doesn't vanish when Marines leave the Corps. "Once a Marine, always a Marine," is a frequently-repeated aphorism, if not mantra, in Marine culture.

Because of the importance of attaining the title "Marine" and being awarded the Eagle, Globe, and Anchor, known as the EGA, the Marine Corps confers no other special status for membership in a specific division, regiment, battalion, company, platoon, squad, and fireteam. This is because you can't subdivide a Marine. No affiliation with a division or other unit is exalted in importance over a "Marine." In the Marine Corps, unlike the Army, you won't see unit patches, ribbons, or badges festooning Marine uniforms, signifying membership in a particular unit. Being a member of a Marine Division does not add or detract from being a Marine. To a Marine, it's irrelevant—and it's also irrelevant to Marine enemies. The only important question is whether he is a Marine or not. Period. Furthermore, unlike other military services, recruits and officer candidates in the Marine Corps go through the same basic training as all other Marines, regardless of their military occupational specialty. So just because a Marine wants to be a computer specialist, truck driver, cook, or mechanic, it doesn't matter: he must still go through the same training as all other Marines. This also applies to officers. Even though some officer candidates want to become pilots or judge advocates, they must still go through the same basic training and education. Becoming a Marine is earned, not given, and there must be no mistaking that someone is a Marine and has earned the title.

I'd be remiss if I didn't point out some exceptions—or at least some circumstances—that collide with the philosophy of "You Can't Subdivide a Marine." First, in the Marine Air Wing, which traditionally is viewed in a different light than Marine ground combat units, Marine squadrons produce and distribute unit patches to its Marine members. But it's important to note that similar such patches exist in most, if not all, of the world's military air forces. Perhaps this is why the Marine Corps distributes patches to Marine Air Wing members for placement on flight suits, not regular Marine uniforms. Second, within Marine ground combat units, there are two elite units: Recon and Force Recon. Although these units, each known generally as an "elite within an elite," confer a special elite status on their members above being a "Marine," they do not distribute unit patches, ribbons, or badges denoting membership. Other "elite-within-elite" organizations within the Marine Corps include the The President's Own (the Marine Corps' Band that plays for the President) and the Silent Drill Team, one of the best, if not the best, close-order military drill teams in the world. Finally, when a Marine gets trained in other US military service schools, like the US Army's Ranger and Airborne (parachute jumping) schools, and the Navy's Scuba school, he is entitled to wear the Ranger patch, Jump Wings, and Scuba device awarded to all graduates of those schools.

6

Marine Difference

Marine culture is wildly different from other cultures, even military cultures. George F. Will, the political pundit and syndicated columnist, in an article on Marine officer training, called the Marine Corps "the military's counterculture." Well, Marine culture is counter all right, but it's also revered and cherished by Marines and former Marines. As a matter of fact, being different and "counter" is partly why Marines and former Marines revere and cherish it so much. From 1775 on, Marines have taken pride in being different. Different from the Army. Different from the Navy. Different from the Air Force. And different from all other military forces across the world. Their worldview is shaped by this perception. They may look like Army soldiers, be equipped like Army soldiers, and sound like Army soldiers, but they are not Army soldiers. And you better not refer to them as Army soldiers. To a Marine, it's a mortal insult. To a Marine, there is nothing "Army" about him, except maybe his weapons and equipment.

Richard Miniter, an author whose father was a Marine, recounted a conversation he had in Iraq with an Army officer about the difference between the U.S. Marines and U.S. Army:

> I had an interesting conversation in Baghdad
> once with a guy named Major Hernandez
> who was a West Point graduate. He said to

me, "Why is it that someone who serves in the Marine Corps for one term has a bumper sticker on his car and is proud to be a Marine for the rest of his life where someone who graduates from West Point never has an 'I love the Army' bumper sticker or flag outside his house?" And I think it's because the Marines are just different. They—it's a big part of their identity, and certainly a big part of my father's identity.

Now, with this Marine pride comes feelings of excellence, greatness, even superiority. Truth is, deep down many Marines feel they are unrivaled. They believe in their supremacy and that other military forces can't compare with them. Many are immodest, if not cocky, about it—and they are also, at least in military terms, probably right in believing in their superiority. Put simply, their battle record is hard to quarrel with. When standing back and looking at Marines, and their feelings of pride and superiority, a case may be made that these feelings contribute significantly to making Marines what they are. Lt. Gen. Victor H. Krulak, in his book *First To Fight*, recounts a conversation he had in 1935 with Gunnery Sergeant Walter Holzworth. Then-Lieutenant Krulak asked then-Gunny Holzworth how the Marine Corps came by its reputation as one of the world's finest fighting organizations. "Well, lieutenant," the Gunny answered, "they started right out telling everybody how great they were. Pretty soon they got to believing it themselves. And they have been busy ever since proving they were right." Lest you think this only occurred in the

"Old Corps" and doesn't occur in today's Corps. Please consider the following boastful exchange, detailed in Robert Kaplan's book, *Imperial Grunts*:

> When I asked a young female sergeant at Camp Lejeune, a pale and innocent-looking wisp of a girl from Kentucky, why she had joined the Marines, not hesitating half a second even, she barked back: "Because they're the best. And I wanted to be the best." Next to her stood a towering sergeant major, an old-timer who described himself as a "southern redneck." He told me, "You know why we're going to win eventually in Iraq? I'll tell ya. Because Marines are there. And Marines don't fail. We just don't. Because from boot camp on up, we learn and relearn our history and tradition.

Lt. Gen. Krulak believes that an elitist spirit and being an amalgam, i.e., part Soldier and part Sailor, goes toward explaining the Marine difference:

> Woven through [the Marines'] sense of belonging, like a steel thread, is an elitist spirit. Marines are convinced that, being few in number, they are selective, better, and, above all, different. This matter of being different has been nourished, over the years, by the Marines' combining of the characteristics of both the sailor and the soldier while not being fairly described as

either one or the other. They have always been alert for opportunities to exploit this anomaly to their own benefit, playing the Army and the Navy against each other.

According to Krulak, the Marine difference also applies to uniforms:

> The determination to be different also manifested itself early on in terms of appearance. Turning to Europe for a model, Commandant Wharton created in 1804 a distinctive Marine Corps uniform, using the already traditional Marine colors of blue, red, and white and including a tall leather cap bearing, as today, an eagle and an anchor. (The globe came later.)

This cultural difference—this "Marine difference"—and the ingrained pride and feelings of superiority, become more apparent when Marines compare themselves with a non-military entity, like American society. To many Marines, Marine culture is worlds' apart from the society which sanctions them. As one might expect, Marine pride shoots way up when they compare themselves with civilians, creating a whiff (even a gale) of hubris in some Marines. Although Marines have their roots in civilian society, they have come to regard civilians as undisciplined and indolent; in Marine culture, the word "civilian" does not have positive connotations. Says George F. Will:

> The making of a Marine...amounts to a studied secession from the ethos of

contemporary America. The Corps is content to be called an island of selflessness in a sea of selfishness, and to be defined by the moral distance between it and a society that is increasingly a stranger to the rigors of self-denial.

Moral distance? It is measured with an odometer, not a micrometer. On the other hand, Thomas Ricks, in *Making the Corps*, believes that the Marine view of American society as selfish, undisciplined, and fragmented is flawed:

> not for what it sees but for what it doesn't see. The Japanese are continually surprised by America's resilience because they don't see the strengths of America—its diversity, flexibility, openness, and inventiveness. James Fallows once made an observation about the Japanese that can also be applied to the U.S. military: "They underestimate the United States," he wrote, "because so many things that can mean vitality in America—immigration, rapid political change—look like chaos to them." Despite its emphasis on values, the American military may be undervaluing America.

7

Marine Stereotype

Many people think of Marines as unthinking automatons, blindly following orders, screaming "Yes, Sir!" all the time, charging up hills and onto beaches with fixed bayonets, being nothing more than muscular bullet magnets. They also frequently picture Marines as dressing, walking, talking, and even thinking alike—or not thinking at all. Of course, that's the stereotype. And frankly, it's understandable. Stereotypes are common and easy to create. But stereotypes are frequently wrong. By swinging a broad brush, the stereotyper avoids critical thinking: it is perilous to argue from the particular to the general. Just because some people act a certain way, doesn't mean all people act that way. While the Marine system is rigid and uniform, and while it produces a remarkably standardized product, not all Marines think alike, lead alike, or accomplish a mission alike. In fact, the opposite is true. Marines have enormous discretion in thinking, leading, and accomplishing missions. This is a surprising fact to many civilians (and also to newly minted Marines). Marine culture, down to its very core, preaches the independence of ideas and solutions. It doesn't want textbook solutions. It doesn't want fidelity to plans when circumstances have changed. It wants its Marines to bloom where planted. It wants them to create. To improvise. To adapt. To take decisions in a turbulent, chaotic, high-pressure situation. And to accomplish the mission in the least costly, most expeditious manner. After all, when you think about it, the Marine Corps didn't establish its incredible battle record (and reputation) by having unthinking automatons applying stale textbook solutions. And Marines are deployed for a variety of purposes: to fight conventional battles, to fight unconventional counterinsurgency warfare, to provide peacekeeping services, to provide humanitarian assistance, and to provide riot control. War is chaos, confusion, and expecting the unexpected. The Marine Corps knows this and gears its training to produce people who can thrive in that environment. Not just thrive, but win. In training, contrary to its stereotype, the Marine Corps continually admonishes its Marines: "It's easy to be hard, but hard to be smart." Smart, tough fighters are winners. Dumb, unthinking automatons are not. It's that simple. In *The Savage Wars of Peace*, author Max Boot had this to say about the Marine stereotype: "Although sometimes caricatured as homicidal Neanderthals (see, e.g., *Full Metal Jacket*), the marines have shown themselves to be the most intellectually supple of the services." Tom Clancy wrote this about Marines: "And some would tell us that Marines are dumb? Like a fox."

Another stereotype is that Marines have no individuality in their personalities and are, in essence, carbon copies of each other. While Marines wear the same uniforms and are expected to follow the same orders, their individual personality (and command style) is allowed to flourish, not be suppressed. *The Marine Officer's Guide*, the "bible" for Marine officers, says this about the Marine Corps' fostering of individuality:

> The Marine Corps cherishes the individuality
> of its members, and although sternly

consecrated to discipline, has cheerfully sheltered a legion of nonconformists, flamboyant individuals, and irradiant personalities. It is a perennial prediction that colorful characters are about to vanish from the Corps. They never have, and never will. No Marine need fear that the mass will ever absorb the man.

In the Marine Corps, when talk centers on accomplishing missions, many questions are raised, like "What are my priorities? Safety? Conservation of weapons and ammunition? Speed? Accomplishing the mission?" Now, this may surprise some, especially civilians, but the number one priority is always accomplishment of the mission. Not safety. Not the conservation of ammo and equipment. Very early on, Marines learn that they are expendable resources, like a piece of equipment. (Marines are, of course, the Marine Corps' most cherished resource, but they are still expendable.) Marines—both officer and enlisted—know this, and know that Marines will be put in harm's way and possibly die to accomplish a mission. While the safety of Marines is important, and can never be forgotten, the mission is always paramount. No matter how bloody it gets. No matter how much equipment is expended. The Marine Corps was constructed on this bedrock principle. And on the principle that there is nothing to be feared more than a Marine rifleman.

8

Attention to Detail

Throughout their history, Marines have learned that ignoring details, even the smallest ones, can lead to dreadful consequences: Marines getting wounded, maimed, and killed—or taken prisoner. Obviously, the stakes in the military are high. Much higher than in the civilian world. As a result, attention to detail in the military is a must. A well-known saying is "The devil is in the details." While this may be true, if you believe in the devil, the devil is not alone. He's in there with genius. Both the genius and devil reside in the details, smiling, waiting for people to miss them, not pay attention to them. Marine culture preaches the religion of rooting out the genius and the devil. Consequently, and intentionally, it creates a culture of perfectionism. But it's not the perfectionism and attention to detail that's surprising, it's the intensity of focus that is truly amazing. For example, everything in the Marine Corps, even the smallest detail, is focused on with such precision it almost always becomes a big deal: rust on a weapon, hair too long, a few pounds over weight, Irish pendant (a small thread) on a uniform, unkempt appearance, being slightly discourteous, forgetting to salute a senior officer, being seconds late. In fact, there is an unspoken presumption in the Marine Corps: Everything is a big deal, unless told otherwise. This presumption, as you might expect, causes many Marines to worry about the smallest, most inane things. But the word "worry" doesn't quite capture it. Marines stress. They sweat. They run here and there asking others how to handle terribly minute matters. They worry so much, and become so "wrapped around the axle," that they eventually become "anal retentive" and turn into "stress grenades"—three frequently repeated phrases in Marine culture. And it doesn't take too long to become all three; attending recruit or officer training will do it. How not to stress? The best way, although not always successful, is to defeat the presumption by having someone in charge tell Marines not to stress. Instructors and senior officers will, at times, qualify the end of their instruction with the following: "Now Devildogs, don't get wrapped around the axle about this." They have to qualify it like this or Marines *will* get wrapped around the axle. It becomes part of their nature. But focusing on the details with intensity, and getting wrapped around the axle, eliminates the devil and lets loose the genius. Look at people who know this. Look at a non-military example: Bob Knight, the former basketball coach of Indiana University and Texas Tech. During a game, he'd counsel a player—read: yell, scream, berate, throw a chair or two—who made the tiniest mistake, imperceptible to viewers, even when the team was up by 30 points with only a minute left in the game. The mistake, of course, had no way to affect the outcome of the game. But that wasn't the point, either to Bob Knight or to any other perfectionist. That same mistake, if left unattended and not worried about, could affect the outcome of a future game, affect the team, affect the school, and ultimately affect—if the mistakes led to losses and to a losing season—revenue. If a mistake is left unattended, the devil will be hiding, lying in wait. So you have to root him out, eliminate him, and find the genius and allow him to escape and flourish. And when you think about behavior like this, it's typical of people who excel. The very best people in their fields—movie directors, musicians, writers, doctors, lawyers, pilots, athletes—all pay furious attention to the details. They know the genius and the devil are in there. Consider the non-military example of commercial airline pilots. In the flying business, a small mistake can lead to catastrophic consequences. And that's why airline pilots are trained to pay attention to the smallest details, and to check and re-check those details. That's why pre-flight checklists came into existence: to pay methodical attention to the smallest details to avoid the needless destruction of innocent lives. In the battle business, it's the same way: a Marine who ignores the details may end up dead, or, even worse, cause other Marines, and even civilian noncombatants, to be injured, maimed, and killed.

During The Basic School (TBS) at Quantico, my company went to the field to play war games. As always, we brought our M-16A2s, helmets, flak vests, deuce gear, and other sundry items. On this occasion, as on most occasions, we used blank rounds, instead of live rounds, to fire our M-16A2s, SAWs, and M-60E3 machine guns. To fire blank rounds, you need a "BFA" to screw into the muzzle. A BFA is a Blank Firing Adapter, a bright red, hollow three-sided metal device with a T-type turn-screw running through it. This device stops the blank round's propellant gases from exiting the muzzle, which forces the gases back to the breech to push the bolt back and chamber a new round.

Well, during this particular field exercise, I realized once I got to the field that I had forgotten my BFA. My bright red BFA. Now, please understand: I brought every other required item besides my BFA. I brought ammo. I brought cammie paint. I brought a compass. I brought ear plugs. And on and on. But no BFA. So you might think, "No big deal, you have everything else, just don't fire blanks. No biggie." Wrong. In Marine culture, it is a biggie. You feel like a schmuck, even for forgetting such a relatively innocuous device. Once other Marines saw that my muzzle was missing the bright red BFA, they asked, "Where's your BFA?" I pensively said, "I forgot it." In response, some said: "You forgot it? Oh." They stood there, silently drawing adverse inferences from my forgetful (and dreadful) act. Forgetting in Marine culture was not good. It demonstrated a lack of judgment and responsibility, regardless of how small and insignificant the item. Other Marines were thinking, What would he do in battle? Forget to bring ammunition? Maps? Radio codes? To them, forgetting in the unforgiving environment of battle killed men. Because of this, when you forgot (or lost or damaged) something, whatever it was, you felt like a loser. You never wanted to let your fellow Marines down. You never wanted them to think ill of you. You never wanted to be different from them, to stick out. At the time of the BFA incident, my personal life really didn't matter. I only wanted my BFA. I only wanted my fellow Marines to forget or overlook my indiscretion. Only in Marine culture could you feel this bad over such a small incident.

9

Selflessness

One overriding trait of Marine culture is selflessness. *Webster's Ninth New Collegiate Dictionary* defines "selflessness" as: "Having no concern for self; unselfish." Marine culture gives a more severe meaning to the term. To Marines, selflessness means putting their bodies in harm's way: for country, Corps, the mission, and other Marines. Using an extreme example, selflessness is what causes one Marine to intentionally smother a grenade with his body to protect the life of another Marine, as Medal of Honor recipient Corporal Jason Dunham did in Iraq, using his Kevlar helmet to smother the grenade and contain the force of the explosion. Think about this. How many civilians would do that for a neighbor or co-worker? Unlike American society, the Marine Corps is not about me, me, me; it's not about individuality; it's not about saving your own life, when other lives are endangered. Instead, it's about devoting and sacrificing yourself to the mission, the unit, the Corps, and other Marines. This aversion of "me-ism"—to include a considerable aversion of "civilians"—starts the minute a recruit or candidate reports to boot camp or OCS. And it starts subtly, linguistically. Upon reporting to boot camp or OCS, recruits and candidates are ordered to eliminate first-person pronouns from their language and to use third-person pronouns. They learn not to ask, "Sir, may I go to the head," but instead to ask, "Sir, this candidate requests permission to go to the head." Sure, this is only a small and subtle linguistic shift. But it leads the recruit or candidate—the civilian, really—to think in a new direction, away from me-ism. Perhaps surprisingly, even this small practice is difficult. It is difficult to eliminate first-person pronouns after using them for the first 20-plus years of one's life. It trips people up. And when it does, it brings on the unbridled fury of drill instructors, platoon sergeants, and sergeant instructors. Of course this is precisely what makes recruits and candidates change: the unbridled fury. In no time, after seeing some of the fury, recruits and candidates stop using "I" and "me" and fondly embrace third-person pronouns. They also become more selfless by helping out, putting the mission first, the unit first, and other recruits and candidates first. Because this linguistic assassination occurs early during a stressful period, and because recruits and candidates are never told why, many of them never understand the rationale behind the proscription. No matter. They do as ordered and rely on what their instructors tell them, which is this: "Everything in basic training has a purpose." Only later, sometimes only after leaving the Marine Corps, do they fully understand the purpose.

Once a recruit or candidate graduates boot camp and officer training and becomes a private or commissioned lieutenant, the proscription against using first-person pronouns eventually relaxes into nonexistence. Even so, some privates and lieutenants still use third-person pronouns— "This recruit [or lieutenant] requests permission to…"— out of habit (or fear of the fury). Now, once a private and lieutenant make it into the mystical and mythical Fleet Marine Force, or "fleet" for short, third-person pronouns disappear altogether. But it's not OK, either at an MOS school or TBS or the fleet, to revert to me-ism again. Marines place paramount importance on acting as a team, not as individuals. Selflessness is still the overriding characteristic of Marine culture, wherever you are stationed or wherever you are deployed. It is hard to get around this learned trait, because Marines constantly police themselves—both with words and conduct (otherwise known as peer pressure). Marines even create their own phrases to keep each other in line. When a Marine diverts from selflessness and shows a tiny bit of me-ism—not sharing food, not helping out, not being part of the team—you'll hear, "*Semper I*," instead of *Semper Fi*. Or, "OK, I see selfish one, you're all about yourself." Or, "I see, you just want to be an individual, hungh?" Like most Marine creations, these phrases are simple and they work. More important, they allow Marines to police themselves, be hands-on, and to reform the conduct quickly. Marines are Devildogs and Watchdogs, all in one.

Before long, because of the internal policing, selflessness is embraced everywhere in Marine culture. As an example, Marines don't stop working just because their own tasks are completed. Regardless of the activity—digging a fighting hole, packing gear, cleaning a room, dressing for inspections—Marines pitch in and help those who are slower, who still have unfinished tasks. No questions. No hesitation. Just do it! Marines also—and this may sound funny—grope and pick each other like little monkeys. They reflexively remove Irish pendants (dangling threads) and tiny lint particles from other Marines' clothing; they fix upturned collars; they align rank insignia and belt buckles, and on and on. All without hesitation and, more important, all without asking. All to help *others* look squared-away. Not themselves, but *others*. All this drums in selflessness.

Selflessness shows itself in eating, too. When in fighting holes—the Marine Corps' term for foxholes—Marines learn first to share their food with other Marines before they take a bite. If they don't, they'll hear about it. ("OK, selfish one, only you need to eat, right?") And when it's time for organized chow, don't think it's OK for all Marines to get in line, load their tray, and spoon it in. Some have to wait. Senior Marines have to wait until their troops have eaten first. And it doesn't matter how senior they are. Fireteam leaders who lead only three Marines don't eat until their Marines have received chow and are eating. Likewise with platoon commanders, company commanders, and battalion commanders—right on up the line. Selflessness. Look to your men first. See that the Marine Corps' most cherished asset—the Marine rifleman—was looked after and cared for. Only then can you look to yourself.

10

Discipline

One Marine Corps hallmark is its extreme discipline and the inculcation of this trait into Marines. I am not referring to "discipline" in the sense of punishment or of a particular field of study. I am referring to the orderly conduct and pattern of behavior of Marines. *The Marine Officer's Guide* defines "discipline" as the "prompt and willing responsiveness to orders and [the] unhesitating compliance with regulations." In other words, immediate obedience.

The importance of "discipline" in a military unit cannot be overstated. Its presence or absence can be the difference between fight and flight, life and death, and victory and defeat. The Marine Corps knows this and therefore stresses discipline from the day a recruit or officer candidate arrives. It has, shall we say, an in-your-face method of instilling it. In the end, it inculcates a type of discipline that causes not only an immediate response to a lawful order, but also a behavior that makes a Marine maintain a "military bearing"—walking tall, talking forcefully (but respectfully), looking immaculate. All this makes him go the extra mile when executing orders and achieving missions. Discipline is about all that.

But it's more than that. *The Marine Officer's Guide* says that discipline keeps a unit together:

> [D]iscipline is a matter of people working well together and getting along well together—and, even if there be a lack of harmony among them, discipline is a means of cementing them as a fighting organization.

Cementing? Well, you need pretty darn good cement when the going gets tough in the Marines. In the Marines, unlike in civilian business life, people get wounded, maimed, and killed—sometimes turned into fine pink mists—right in front of your eyes, and on a daily basis. Discipline is the cement that helps keep all the loose and individual components a part of the whole—it's the cement, along with peer pressure and the fear of being ostracized, that keeps a military force cohesive, that produces fight instead of flight. As George Washington said: "[T]he distinction between a well regulated army, and a mob, is the good order and discipline of the first, & the licentious & disorderly behavior of the latter."

The preferred type of discipline is the self-perpetuating type, not the autocratic type, and so the Marine Corps preaches self-discipline. As *The Marine Officer's Guide* states:

> The best discipline is self-discipline. Self-discipline amounts to the Marine having control of himself and doing what is right because he wants to. To be really self-disciplined, a unit must be made up of men who are self-disciplined. In the ultimate test of combat the leader must be able to depend

on his men to do their duty correctly and voluntarily whether anyone is checking on them or not.

The Marine Corps, through its culture, teaches Marines to execute orders and missions in the quickest, smartest, and highest quality way possible. When it doesn't think this is happening, it makes its Marines aware of this fact in short order: senior Marines tell junior Marines that they're behaving like Army soldiers or civilians in the "Civ Div" (Civilian Division), true insults to Marines. This all gives Marines an incentive when executing orders and missions: execute quickly or be ridiculed as being mediocre and slow.

In the end, the Marine Corps inculcates discipline and self-discipline, making its Marines elite: it makes them cohere on the battlefield, it makes them execute lawful orders unhesitatingly, it makes them run toward gunfire and into harm's way, and it makes them go the extra mile to be the best at whatever they are doing: cleaning the squadbay, shining boots, pressing utilities, running races, taking exams, doing whatever is asked of them.

11

Integrity

Marine culture produces Marines who act with copious amounts of integrity—at generally much higher levels than in civilian life. Marines are taught the concept of integrity early on, through classroom instruction, practical application, and real-world experience. There is much pressure to act in accordance with integrity in Marine culture. Over and over, Marines are told to never "lie, cheat, or steal"—just like students at the military service academies. At The Basic School (TBS), the following comments by Major P.B. Johnson about integrity were placed in the "Welcome Aboard" packet given to all new lieutenants of Fox Company:

> My final comment has to do with perhaps one of the most important concepts here, integrity! By virtue of your position and rank only the highest standards of integrity will be accepted. Officers who lie, steal, cheat, etc., have no place leading Marines or in the Corps. Common sense and a high set of values and standards will guide you in this area. Your word must be your bond as a Marine officer if you are to lead the world's finest warriors – United States Marines.

Like the virtue of selflessness, Marine culture produces internal policing mechanisms for integrity. Marines say,

"Integrity Violation," when witnessing something they deem inappropriate. Or "Integrity Check," when encountering a situation requiring an appraisal of integrity. What is integrity? How is it defined? Even though the notion of integrity is drummed into Marines during recruit and officer training, and again in the fleet, many Marines still never understand the precise definition of it—at least not to recite it from memory. Nor are many Marines capable of distinguishing the difference between integrity and honesty. But despite being unable to articulate precise definitions, most Marines figure out how to act with integrity (it's like pornography: they can't define it, but they sure know it when they see it). In fact, when it comes to integrity, Marines are generally much better at it than civilians (perhaps because they focus and stress about everything). For example, something small might happen to a Marine, and then the internal questions start: Is this right? Is it wrong? How should I respond? Is my response right or wrong? Should I tell my superiors?

Perhaps the best definition of "integrity" is contained in a remarkable book titled quite simply, *Integrity*, written by Dr. Stephen L. Carter, a professor at Yale Law School. In his book, Dr. Carter defines "integrity" as having three components. The first is discerning right from wrong; the second is acting in accordance with what had been discerned as right, even if there is a personal cost; and third is saying openly what was discerned as right. This puts teeth in the definition of a term that is frequently viewed as vague and indefinable. Nevertheless, this virtuous quality has its drawbacks in Marine culture.

Marine culture produces Marines who are, in the words of some, "self-righteous" and "Holier than thou." For obvious reasons, some Marines think they are better—more moral, more integral—than other Marines. And some, because of this, "get weird on you." This is where integrity can have unsettling and controversial characteristics. Many Marines have a penchant to tell their command about certain activities, such as someone doing something unethical or illegal, or even something as simple as someone not pulling his own weight or working as hard as others. This proclivity to "tell," and of course to name names, is acute in Marine culture—certainly more acute than in civilian life. And it isn't hidden, either. It is there, in full bloom, for all to see. And it naturally has a chilling effect. It makes you wary. It makes you think twice before you tell a story to another Marine. Fact is, you can tell a story to a Marine, a funny story, and before you know it, you're explaining the whole thing to the command. Why does it happen? What motivates Marines to "tell"? Why is this penchant for "telling" more acute in the military?

The Marine proclivity to "tell," in my view, springs from one or more of five sources. First, it comes from real-world applications of integrity's definition: discerning "right" from "wrong" and acting in accordance with it. Second, it comes from the notion of personal accountability, and fearing the adverse consequences of not saying or doing something if one's superiors find out about it. Third, it comes from a legal duty, in some circumstances, to report unlawful activity. Fourth, it comes from ill-will, a dislike of the person being reported on. And finally, and most unfortunately, it comes from a desire to look better in the eyes of the command or seniors. Whatever the source, in whatever combination, it motivates a Marine to take an affirmative step to "tell." This is not to say that telling is easy, for it typically is not: it causes much internal consternation, struggling, and deliberating. Additionally, the added concept of a friendship—and imperiling it if a decision was taken to tell—makes it even more difficult to tell. When a friendship is involved, a Marine will employ an informal (if not subconscious) balancing test. He will weigh the value of integrity—or one of the four other reasons for "telling"—against the value of a friendship, like it's some sort of virtuous equation. Some do this consciously, some subconsciously. And some try to mitigate the potential damage to the friendship by talking to and trying to convince the friend he was "right," that he had to do it because he had no choice (in essence, making it a win-win situation). Often times, despite being friends, or even close friends, the Marine will tell. He will apply the definition of integrity, or one of the other virtuous reasons, weigh it against the notion of "friendship," and find the scales tipped in favor of telling.

While stationed at Camp Pendleton, California, and living in an apartment in San Clemente, I learned of adulterous activity then occurring between my neighbor's wife, a civilian, and a Marine Lance Corporal stationed at Camp Pendleton. The wife's husband was a Navy Ensign, who happened to be away one day on his ship as Officer of the Day at the Long Beach Naval Base. I observed, in plain view, the Lance Corporal put his arms around my neighbor's wife and kiss her. I saw no further sexual activity, although I was told, from a friend of my neighbor's wife, of such activity occurring many times with this Marine (including other service members).

Fast forward to Monday, two days later. I was driving to work with another Captain, who also happened to be a Judge Advocate at Camp Pendleton, and a good friend. While driving, I told him the whole story, all the details, just like a good little lawyer would tell it. He said it was a terrible story, and that he was going to tell his command—in fact, he was going to tell them as soon as I dropped him off. I thought he was kidding. He was not. I was nonplussed. I told him not to say anything to the command, because what had occurred was a "personal matter" between my neighbors themselves, and it wasn't a serious crime that should be reported, like murder, arson, burglary, or robbery. And furthermore, I told him I had to live next to them. My friend was unpersuaded. Again, I told him not to tell. I even asked him not to tell, friend to friend. He would have none of it. He was telling, end of story, despite my vociferous objection. Agree to disagree. To his credit, he told me what he was going to do, and he told me his reasons: he felt he had a duty to report it, and he wanted the Lance Corporal court-martialed for adultery. After I dropped him off, he went, as promised, straight to his commanding officer, a Lieutenant Colonel, and told him the whole story. All the adulterous details. What eventually happened? Nothing. No non-judicial punishment. No court-martial. Nothing. In the Lieutenant Colonel's opinion, there were not enough facts to prosecute. And he wanted to see whether the Navy Ensign would seek prosecution or at least an investigation. But what ultimately happened is not the point of the story. For me, the point was clear. Marines get motivated, for a variety of reasons, to tell the command about certain activities, and that usually requires naming names. The lesson was clear, too. Be wary who you tell, even friends.

Was my friend wrong? Was this proclivity for telling—and my friend's actions—the same as "ratting"? Of course, it depends on how you define "rat." Webster's Ninth New Collegiate Dictionary defines "rat" (in addition to being a four-legged furry rodent), as one who betrays, deserts or informs on one's associates. This is a broad definition. Much depends on how "associate" is defined, too ("partner, colleague, companion, comrade"). Thus in my example, my friend is probably not a rat, for he was not informing on one of his associates (that term implies "friendship"). But there is another definition of "rat," supplied by G. Gordon Liddy, a man who knows a few things about rats. During Watergate he was, and still is, the quintessential anti-rat. He defines a "rat" as someone engaged in a criminal enterprise with others, and who will provide testimony about the enterprise or against people involved in it, in exchange for favorable treatment of any kind. Therefore, to be a "rat" under G. Gordon Liddy's definition, one must be involved in the criminal enterprise and get favorable treatment in exchange for talking about the enterprise or the people. This definition excludes a great many people, like my friend who told his command, and people who are witnesses to criminal activity but who are not involved in it. (The Webster's definition is broader, because it would encompass mere witnesses testifying against associates, even though the witnesses were not involved in the criminal enterprise, nor given favorable treatment.)

Why was the proclivity to tell more acute in Marine culture? Well, once again, Marines focus on every little thing that happens, or doesn't happen. They don't brush things off or forget about them (unlike the tendency in civilian society.) They confront them, employ the balancing test, decide how to deal with it, and take action. Simple as that. By and large, I don't think it's a bad thing. When all is said and done, Marines seem to be motivated to act by virtuous reasons far more often than civilians.

12

Physical Toughness

Marine culture produces and fosters physical toughness. No other way to put it. After all, Marines wage war, and waging war, more than anything else, is a test of wills—one side trying to impose its will on the other, like two Sumo wrestlers, as Karl von Clausewitz analogized in his magisterial book, *On War*. Even though warfare has evolved through the centuries, much of it remains the same. It is still a horrifyingly bloody business, with the killing of human beings at its core. "War means fightin' and fightin' means killing," said Nathan Bedford Forrest. "War is cruelty. And there is no way to refine it," said William Tecumseh Sherman. And war still requires tremendous physical exertion. As Maurice de Saxe said, "Victory goes to the strongest legs." The weather and terrain still exhibit the same unharnessed ruthlessness—all on people who are tired and stressed out to begin with. So building the will still remains key; it pushes the Sumo outside the circle. Although predominately mental, building the will requires intense physical training (known as "PT'ing" in Marine culture). It helps produce the stamina, the toughness, the courage—and ultimately the strong will. Because of this, PT'ing is part of Marine culture. Drive on a Marine Base and you'll see muscular bodies with shaved heads crisscrossing the base, running along roads and trails, doing pull-ups, sit-ups, the whole bit—in far higher numbers than in civilian society. And you'll see them PT'ing not as a company, platoon, or squad, although many times you'll see that, but alone or in groups of two, three, or four (called "Individual PT," even though it sometimes consists of small groups). PT'ing becomes part of a Marine's daily routine, like showering and shaving. In fact, most Marines come to love PT. They learn that PT'ing produces many good things—it builds the will, sheds the weight, stimulates the mind, and releases endorphins which make them tougher, faster, stronger and—much to their delight—younger looking. Consequently, Marines, even without being ordered, PT as much possible. What is more, because of the Sumo and their desire to win, they willingly push the physical envelope to new peaks of pain. George F. Will, in one of his syndicated columns devoted to the Marine Corps, stated that the Marine Corps exhibits a "conscious cultivation of an ethos conducive to producing hard people in a soft age."

But let's not mistake this physical toughness for the absence of fear or nervousness, or better yet, the absence of the outward display of fear or nervousness. Contrary to the civilian bumper stickers saying "No Fear," Marines are taught that feeling fear and showing it through nervousness is absolutely OK. In fact, it's expected. After all, Marines are supposed to go "into harm's way"—a phrase coined by John Paul Jones—to kill people. That endeavor makes most people nervous. But more importantly, what is expected of Marines is that, while feeling fear and showing it, they do not become immobilized by it. That's the thing. It's OK for a Marine to feel fear, to show it, but he better not sit there and do nothing. He better work through it. One foot in front of the other. Move! Move! Move! Marines were told this time and again. It was not uncommon to see Marines displaying nervousness—quivering lips, dry mouths, fidgety hands and legs, being quiet—during all sorts of activities: getting personally inspected in formation; preparing for an arduous PT session; getting called on in class; having to report to the Staff Platoon Commander or Company Commander; assuming a significant leadership billet, and so on. But they knew they better not be immobilized by it or, worse, to run from it. This would get them ostracized even before they could say "ostracized."

Despite this culture, and despite the good things PT'ing does for them, you can still find a few Marines who don't like to PT—and who cut corners. Some turn their dislike into inaction, declining requests to PT made by fellow Marines. These types of Marines are noticed immediately. Few things in Marine culture are more disfavored than non-compliers. And few things are more disfavored than non-compliers who are lazy if not fat. Now, once a Marine declines some requests to PT with others, the internal policing mechanism—peer pressure—kicks in. Peer pressure consists of coaxing and providing humorous but sharp comments. And depending on the circumstances, some Marines even rate the special "intensified peer pressure," which is like regular peer pressure, except that the comments are made publicly in front of others. And the comments become, as more and more are uttered, more sharp and less humorous. In most cases, before you know it, the little non-complier is having a jolly time running, lifting, and eating well.

Another thing disfavored in Marine culture is going to sick-bay or seeking medical treatment—even if warranted and legitimate. Marines who seek medical attention or go to sick bay are quickly branded—depending on the severity of the injury and the person seeking treatment—as a "sick-bay commando," a "non-hacker," or a "weak sister." Some Marines, of course, get branded undeservedly. But truth be told, some Marines are weak and will seek medical attention to avoid the brutal physical exercises that accompany training (getting a "chit," a written authorization, from a doctor legally exempts Marines from physical exercises). Therefore, a reason exists to brand some Marines as sickbay commandos, non-hackers, and weak sisters: it gives credit where credit is due. Aside from being deserved, this branding and peer pressure contributes to a culture of toughness and machismo. Most Marines learn to avoid sick-bay at all costs. They learned to tough it out, no matter what the cost in pain and tears. And it only makes sense not to go to sickbay— not only because you didn't want to be ostracized, but also for another reason: Marines are supposed to be—they must be—tough to accomplish their missions and to survive; they aren't supposed to need sick-bay. If they aren't tough, and need sick bay, they must not be Marines. Marine Corps' history is replete with countless battlefield examples of men triumphing over incredible odds, suffering unimaginable physical and psychological pain, while sometimes even making the supreme sacrifice, all to accomplish the mission. Knowing this, Marines generally don't want to hear whining about physical hardship or pain. It's an insult to their antecedents. Those past Marine sacrifices typically dwarf any predicament a modern-day Marine might find himself in, especially in training. So Marine culture infuses this history into their Marines. In fact, it never lets them forget it. This is because it causes them to act like their antecedents.

Another characteristic of Marine culture, perhaps an unintended characteristic, is the production of macho. It is prevalent in the Corps, as you can well imagine, because its adherents believe, erroneously as we'll see, that a macho attitude is equivalent to toughness. It is not—or at least not in most cases. Macho is fake; toughness real. In many cases, as one learns from life, macho is a thin veneer hiding fear and weakness. Real toughness needs no macho veneer. It's like talk versus action. Talk is cheap (unless of course you hire a lawyer); action and results are what counts. Ernest Spencer, the author of *Welcome to Vietnam, Macho Man*, had a close and personal view of macho as a Marine lieutenant at the battle of Khe Sanh. Here's how he described "macho," which he equated in the end to approval-seeking behavior:

> Being a Marine was all about being a macho man. Macho is not just played by military types, though. Anybody can play macho. Macho can also mean bullshitting yourself and others as much as possible. The problem with playing macho is that macho is about being afraid—afraid of rejection. It is a desire to be loved, wanted, and respected. Being macho is depending on others for your status. A macho guy is the counterpart of the woman who is locked into how she is perceived by others rather than who she herself is. There's not much difference between a macho guy and a woman who keeps trying to find happiness with a senseless jerk rather than in herself.

On Marine Culture

Macho is being afraid to look at yourself and laugh—to laugh for caring so much about what others think. Fear of rejection gets a woman rejection after rejection; it gets a guy killed.

13

Aggressiveness and the "Warrior Spirit"

Marine culture fosters physical and mental aggressiveness, culminating in the inculcation of an offensive mindset and the "warrior spirit." Fact is, Marines don't like to defend, mentally or physically. They like to attack. And when you think about it, there is something energizing about aggressive action and taking the offensive. Not only does it make you think you are in control—calling the shots, making the enemy react to you—it is like vigorous PT: it does something to your brain and psyche. Endorphins are released. Ideas come quickly. In short, aggressive and offensive action emboldens you. On the other hand, on the defense, you're passive, sitting, waiting for the enemy to act, so you can react. Action versus reaction. Offense versus defense. Big difference. Marine culture does not favor inaction, reacting, and being slothful. It favors action, aggressiveness, spirit. "Attack, attack, attack," said General George S. Patton, when he was asked about the principles of war. He could have been a Marine.

"Lead, follow, or get the hell out of the way!" is a frequently repeated phrase in Marine culture. It embodies the aggressive spirit of the Corps. Marines promote and foster decisiveness, and people who are decisive. But in Marine culture, it is not good enough just to be decisive, you have to be quickly decisive. You can't sit on things. You can't dawdle. Inaction is the worst thing in Marine culture. It gets officers relieved just like that. Marine culture emphasizes getting the facts, analyzing them, and taking a decision. Right then. Not later. Marines are admonished that not taking a decision and sitting on it is actually taking a decision—to sit on it. Warfare has produced enough examples of commanders who have dawdled and didn't take decisions quickly enough—all to the detriment of their troops and nation. Civil War Union General George B. McClellan probably takes the cake—Lincoln even described him as having "a bad case of the slows." In any event, delaying decisions causes missed opportunities, excessive casualties, bad morale, and ultimately defeat—in body and spirit.

To combat this, the Marine Corps teaches a theory called the "OODA Loop." This theory, created by iconoclast Air Force Colonel John Boyd, specifically encapsulates the decision-making process that a fighter pilot must go through to prevail in a dogfight against an enemy pilot: he must Observe, Orient, Decide, and Act. In other words, he must be the first one to see, gather information, take a decision, and then put the decision into action. According to John Boyd, a pilot who performs these OODA Loop actions the quickest is the pilot who prevails—nothing more, nothing less. The Marine Corps adopted this OODA Loop theory not only for its fighter pilots, but also for its Marines generally. The beauty of the theory is that it is general enough to be applied to virtually any activity. In fact, many businesses have now adopted the theory to help them shorten their decision-making cycles and to prevail against their competitors. As one might expect, the Marine Corps placed paramount importance on teaching the OODA Loop theory to its officers, and in having its steps executed with alacrity. As an indication of how extensively the OODA Loop theory permeated Marine culture, Marines frequently used the phrase "You're in my loop" to other Marines who, it so happened, were getting the best of them in one way or another.

All this produces a feeling in Marines tantamount to a constant hot breath on the back of their necks—making them feel like they have to be aggressive, like they have to continually go, go, go, and to decide, decide, decide, as if someone was always right behind them, watching them, waiting to take advantage of their missteps, their inaction, and their indecisiveness.

14

Barking

Part of the Marine culture of toughness is "posturing." In his book, *On Killing*, Lt Col Dave Grossman (USA) describes "posturing" as part of the "fight-or-flight" psychological model for military people. He defines it as physical actions taken by a person to harmlessly intimidate others—through sight or sound—before fighting or fleeing. We see this in military history, he says, when warriors try to psyche out their enemy immediately before battle. He gives examples: the "rebel yell" by southerners during the American Civil War, the whistles and bugles of the Chinese and North Koreans in the Korean War, and the "hurrah!" of the Russian infantry. Well, surprise, surprise, this posturing is also present in Marine culture. But Marines don't yell, blow bugles, or shout "hurrah!" They bark. Matter of fact, they love to bark. After all, dogs love to bark and Marines are known, courtesy of the German Army in WWI, as *Teufelhunden* or Devildogs.

Marines learn to bark about the same time they learn to salute, which is early. Now it's hard to capture with the written word how Marines bark. It's not woof, woof. It's something like this: OOH RAAHHH! Now, that's how Marines bark on paper. When it's performed live and in-color, it's a different thing entirely. The bark is loud and originates, if done properly, deep within the belly (it's not tinny or shallow). It's performed with the whole body and spirit and is almost violent when rendered by a Marine skilled in the art. First, his eyes open full-bore, fill with fire, fix on yours; then his arms and torso stiffen, his mouth opens. Then, suddenly, finally, in what amounts to a gyration like swallowing air for a burp, it's unleashed with a quick fury: OOH RAAHHH! Of course, one needs an in-person demonstration to fully appreciate the Marine bark.

It is important to know that Marines bark all the time, even in peacetime, and that barking is not always intended to intimidate others. It is a kind of all-purpose device, to be used on many occasions, for many things, in many locations. As Robert Kaplan wrote in *Imperial Grunts*, "Ooh-rah, like Hoo-ah [the Army phrase], meant roughly the same thing: 'Roger.' 'Great.' 'Good-to-go.' 'How ya doing?' 'Stay motivated.'" Thus, it is used as a greeting for Marines. But it's also used for sadistic things and masochistic things. Things showing the macabre Marine humor. So Marines will bark at physical pain (like an upcoming, grueling PT event), things demonstrating toughness, adversity, killing, maiming. Yes, I know it sounds weird. But that's the way it is: a Marine is taught to love and bark at anything that produces pain, killing, or maiming. He also loves to bark at any reference to the Marine Corps or to its illustrious history.

Now as much as it would be easy to say that Marines exclusively use "ooh rahhh" as their phrase of barking choice, it would not be entirely true.

My Staff Platoon Commander at The Basic School (TBS) substituted the words "bark, bark" for "ooh rahhh!" He

On Marine Culture

would say "bark, bark" in a normal tone of voice in lieu of the deep, violent, loud rendition of "ooh rahhh!" He would end platoon meetings, inspections, et cetera, with the quiet but very funny "bark, bark."

Bark, bark.

15

Teaching Methods

The way in which Marines are taught reveals some of the reasons for their success. They are taught using a combination of three principal methods: classroom instruction, sand table exercises, and practical application ("Prac Ap") exercises. Although these methods usually proceed sequentially from classroom instruction, to sand table exercises, and to "Prac Ap" exercises, this is not always the case. Sometimes the subjects begin and end in the classroom—or in the field. The subjects taught are generally called "packages," which, to name some, include land navigation, patrolling, platoon attacks, physical training events, defensive tactics, offensive tactics, artillery call-for-fire, helicopter operations, rifle and pistol qualification, weapons' practicals, weapons' live fires, and Marine Corps history and tradition. Usually the course of instruction blends the different "packages," in which a recruit's or candidate's main worry shifts from week to week. Each "package" storms into their poor lives two to three times a week, for a few hours, and then storms out. Thus each week contains an admixture of packages, with a package lasting anywhere from one to eight hours a day, and from one to a few months overall.

The Marine Corps, on balance, has extremely good instructors. And when you think about it, it's logical. Besides engaging in battle, the military is all about training people. Thousands and thousands of people. Every day of the week. So training is ingrained in the Marine Corps—and in

Marines. Now, to be sure, the instructors, who vary in rank, are at times autocratic, uncaring, and loud. *"Lieutenant, did you see what you did with that slick maneuver?"...You killed all your men!...Every last one of them!...Beautiful!...Do you have a death wish or what?* But for the most part, the instructors are exceedingly competent, passionate, remarkably effective teachers, who sincerely and sometimes desperately want Marines to learn and comprehend the subjects being taught—which usually began and ended with tactics: squad, platoon, offensive, defensive, and weapons.

Many Marines, usually the infantry-types, hate classroom instruction and want to be "in the field," a euphemism for being outside, in the dirt, playing war.

I liked attending class. Each of the four main classrooms in TBS's Heywood Hall at Quantico had a movie screen on the forward wall. Once we were seated for instruction, and sometimes even as we entered the classroom, there would be a violent movie clip playing, or real footage from an air raid during a war. No sound would come from the movie. Only thing you heard, and believe me it was loud, was blaring rock & roll, like Van Halen and Guns & Roses and Motley Crue. Commingled with the music were lieutenants ooohh-raahhing at the sight of bombs bursting, people dying, and buildings disintegrating. Macabre Marine humor. The video and music were called Attention Getters, and they worked. It was, by far, the most enjoyable part of classes. Some instructors took it a step further. After the short video, some would grab a book on Congressional Medal of Honor recipients and read battle accounts where the CMH was awarded. At the end of the recital, lieutenants oohh-raahhed without interruption. And then the class began.

Teaching Methods

For learning tactics, the teaching progressed in a logical way at TBS. We first participated in a STEX, then a TEWT, and finally a FEX. A STEX? TEWT? FEX?

A STEX is a Sand Table EXercise. STEXes were only taught to us in classroom 4. In the middle of CR4 stood a huge rectangular sand table, surrounded on three sides by aluminum bleachers. The tactics were first introduced to the company on this big sand table. Then we broke into groups of 8-12. These small groups went to the perimeter of the classroom, where there were approximately 20 smaller sand tables. A Captain or a Major led these more manageable groups in a free-wheeling tactical discussion, a tactical scenario. The instructor wanted—demanded—participation from all the students. In fact, he usually picked three of us to assume the role of company commander, platoon commander, and squad leader. Then we got to play in the sand. We moved plastic men, material, and weapons to various spots, until we were happy. We placed these figures in positions we thought advantageous, given the tactical scenario before us. We had to explain why. Then we got critiqued. After the critique, we moved outside to Training-Area 8 (TA-8), the area surrounding TBS-proper, to participate in a TEWT.

A TEWT is really a misnomer. It stands for Tactical Exercise Without Troops. But, you see, there were troops. We were the troops. And believe me, contrary to our desires, we were there and participating in the tactical exercise. A TEWT seeks to accomplish putting newly taught tactical concepts to use by— and this phrase gets worn out by the end of TBS— "walking the dirt" or "walking the dog." One must "walk the dirt," get a feel for the lay of the land, the undulations, the denseness of the forest, to effectively employ the equipment and Marines provided him.

The TEWT, as with the STEX, was usually taught by a Captain or Major. He explained the situation to the group—usually platoon-sized, but sometimes squad-sized—then allowed us to wander around the dirt, deciding where we would specifically employ our SAWs, M-60E3s, M-203s, and fireteams. We usually were assigned the role of squad leader, but at times we were the platoon commander. After wandering and deciding on specific locations for men and weapons, we returned to the instructor (all officer instructors were called AIs, for Assistant Instructors; all enlisted instructors were EIs), who would then pick one of us to present a detailed plan to the platoon. Spears and arrows were launched. All the other lieutenants, not to mention the AI, got to critique the poor guy's plan. Most of the AIs were courteous and wouldn't excoriate him for having a bad idea. To be sure, there were some instructors who would take egotistical pleasure in demeaning a new lieutenant. But the clear majority honestly believed there were no "right answers"—as long the plan was backed with a reasoned and logical analysis.

A FEX is a Field EXercise. In the learning progression, it followed both the STEX and TEWT. In short, it was a deployment to the field to employ tactical concepts, such as offensive and defensive tactics, patrolling, convoy operations, attacking a fortified position, and so on. Some FEXes lasted only a day; others lasted two-, three-, and four-days, with the 9-Day War as the epoch. If we went on a 1-day FEX, we returned to the BOQ anywhere from 1500-1800 hrs. If we went on a 2-, 3-, or 4-day FEX, we stayed overnight in the field. Our days in the field usually ended around 1700 hrs, whereupon tents were pitched, fires started, and stories of the day recounted, modified, and glorified.

Teaching Methods

For most FEXes we had to draw weapons—SAWs, M-203s, M-60E3s, SMAWs—from the armory. We also had to draw binoculars, night-vision scopes/goggles, and radios. But the drawing part wasn't so bad. It was the returning part which tested our patience. The weapons, after dragging them through the red mud of Virginia and firing countless blank rounds through them, were filthy. Filthy beyond description. It also didn't help to have the BFAs screwed into the muzzles of our weapons. BFAs only added to our misery. They stopped all the gases from the blank rounds from leaving the muzzle, which turned the insides a nice shade of black. BFAs acted as a cork at the end of the barrel and were a necessity for gas-operated semi-automatic or automatic weapons firing blank rounds. So as you can imagine, thanks to the red mud, blank rounds, and BFAs, our weapons were black. Blacker than night. Cleaning them, therefore, was never quick and easy. And trust me, we weren't going anywhere until that armorer—usually a lance corporal— put his tiny initials on a card, signifying that the weapon was clean and OK to be returned to the armory. We hated this. Frequently our entire company would return in the late evening from a two-, three-, or even four-day FEX on a Thursday or Friday night. But before any lieutenant could get secured for the night, all the weapons checked-out from the armory had to be returned. Cleaned. And spotless. So when we made plans before the FEX, we were wary. It always required an extra three hours at the armory once we returned from the field. And we never knew the exact time when we would return from the field. Because of this, many lieutenants would "conserve" their ammo by not firing their weapons; they didn't want that powder piling up, causing them to scrub, scrub, scrub. They wanted tiny little initials on their cards as quickly as possible.

The majority of FEXes were platoon tactics—both offensive and defensive. One of the first FEXes was "Squad-a-Thon," a two-day evolution employing a squad in the offensive. For this FEX, we were picked randomly to play the role of squad leaders, to receive and issue a 5-paragraph order, and to lead our 13-man squad to the objective. This exercise was similar to SULE II (Small Unit Leadership Exercise II) at OCS, except it wasn't graded and weighted—neither were any of the other FEXes, TEWTs, and STEXes—and there was no tortuous 15-mile hike to get to the training area. Almost every one of us was picked to lead the squad over the two days. Only one or two of 13 lieutenants were successful in dodging the icy glare and pointing finger of the AIs. (The TEWTs were a little different: only infrequently did you get "volunteered"; mostly it was the other guy getting "volunteered.") All the FEXes worked this way. Except that FEXes involved leading more than just squads. They usually required leading platoons of 45 men or patrols of 15.

Did I mention terrain models? Before every patrol and platoon FEX, when we first arrived at the training-area, we had to build a terrain model—whether we liked it or not. The student squad leader (or platoon commander or patrol leader) picked another lieutenant to help him. A terrain model is an area on the ground that is boxed-out (approximately 8'x10') and accurately reflects, on a much smaller scale, the terrain the unit would be passing through to get to the objective. Hills were shown by piling dirt or leaves or snow. Rivers were shown by using blue yarn or blue spray paint. Units and "attack points" and "assembly areas" and "lines of departure" and "objectives" were shown by writing on 3-by-5 cards. Draws and valleys were shown by digging out some dirt. When the unlucky lieutenant studied his map long enough and put the notable features inside the terrain model, he issued the 5-paragraph order (or patrol order) to his men while standing over this model. He walked through it, pointing and touching different areas for emphasis.

16

Embracing Ambiguity

Marine culture teaches its Marines to embrace ambiguity. They teach it straight on—in classes, in school circles in the field, and in off-the-cuff conversations between seniors and juniors. They tell you that, in battle, you'll never have enough information before taking a decision. You'll always want more—more facts, more information, more guidance. They teach you that the people who are decisive and quickly take decisions in a chaotic, turbulent, and ambiguous environment, will generally prevail against their opponents. So the Marine Corps teaches its Marines to embrace ambiguity. For instance, in sand table exercises, a handful of Marines will gather around a small sand table, approximately the size of an air hockey table, filled with small tanks, artillery, mortars, machine guns, and the like. The Instructor will give the Marines a scenario, pick a leader, ask the leader, in conjunction with the other Marines, to emplace all the weapons and equipment in the best positions possible, and then have the leader explain why he placed the forces, weapons, and equipment in those positions. Frequently, the Instructor will change the scenario by adding or subtracting enemy (and friendly) forces, weapons, and equipment. The Instructor will put additional pressure on the leader by telling him he only has 10 seconds left, and that for every additional second he takes to decide, that one, two, three of his Marines are getting wounded, maimed, and killed. This also happens on a larger scale, on a bigger sand table, or on a field exercise, in front of a company of Marines, where Instructors *volunteer* a Marine to come up in front of the company of 250 Marines to endure fast-paced questioning and provide fast-paced answers.

17

Shooters!

In addition to their physical conditioning, discipline, and warrior spirit, Marines have another attribute that distinguishes them across the globe: their marksmanship abilities. Quite simply, wherever and whenever Marines are deployed and put into harm's way, they consistently field the best shooters. The Marine Corps treats marksmanship as sacrosanct, instilling it in every Marine. In fact, "Every Marine a rifleman" is their credo, and qualifying on the rifle and pistol ranges, as you might expect, is a long and arduous endeavor. To show you how crazy Marines are about their love of rifles and marksmanship, just consider the words in *My Rifle: The Creed of a United States Marine*, created in 1942 by Major General William H. Rupertus, USMC, a sonnet taught to all Marines during recruit and officer training:

> THIS IS MY RIFLE. There are many like it but this one is mine. My rifle is my best friend. It is my life. I must master it as I master my life.
>
> My rifle, without me, is useless. Without my rifle, I am useless. I must fire my rifle true. I must shoot straighter than any enemy who is trying to kill me. I must shoot him before he shoots me. I will…

My rifle and myself know that what counts in this war is not the rounds we fire, the noise of our burst, nor the smoke we make. We know that it is the hits that count. We will hit . . .

My rifle is human, even as I, because it is my life. Thus, I will learn it as a brother. I will learn its weakness, its strength, its parts, its accessories, its sights, and its barrel. I will keep my rifle clean and ready, even as I am clean and ready. We will become part of each other. We will . . .

Before God I swear this creed. My rifle and myself are the defenders of my country. We are the masters of our enemy. We are the saviors of my life.

So be it, until victory is America's and there is no enemy, but Peace.

When I was at Quantico, we had "Rifle Qual" at TBS. This was the rifle qualification course, known as the "KD Course," or Known Distance Course. At the time, it was known as the most difficult military rifle qualification course in the world. Using an M-16A2, a Marine had to shoot at 200-, 300-, and 500-yard targets, all without the benefit of telescopic sights. Only the standard front- and rear-sights provided on the M-16A2 were allowed. And let me tell you, 500 yards is a long, long way to successfully engage and hit a target. But the Marine Corps taught us in exquisite detail how to properly breathe, squeeze the trigger, and adjust the M-16A2's sights for a given distance, elevation, and windage, and to hit and kill the enemy.

It was a two-week, total time commitment, which also included pistol qualification.

The pistol qualification course, although not as revered and sacrosanct as the rifle qualification course, taught us how to use a Beretta 92F 9mm pistol safely and accurately. The pistol range used bull's eye targets at 25 yards, 15 yards, and 7 feet. In addition to untimed firing, the course had a timed firing component, testing our reloading and aiming speed by making us hold the pistol "at the ready" (holding it with both hands at a downward 45-degree angle) and then raising it up to get sight alignment, firing two shots, dumping the magazine, reloading, realigning, and then firing two more rounds.

There were three ratings and badges possible for Marine rifle and pistol qualification (from best to worst): Expert, Sharpshooter, and Marksman. It was a high honor in the Marine Corps to shoot and rate an Expert badge. This is because shooting well is one of the most important parts of being a Marine. A Marine garners enormous respect from other Marines if he can shoot the lights out, even if he's not a particularly good leader or good Marine. Marines know that when the chips are down, this Marine, more than any other thing, would be able to kill the enemy well, save Marine lives, and help accomplish the mission. To a Marine, not much else matters.

18

Love?

In Marine culture, despite being a culture of toughness and machismo, Marines love one another. Although same-sex love sounds unbelievable in the macho—and some would say homosexually-averse—environment of the Marine Corps, it is true. The bonds formed between Marines are sometimes stronger than the ones between a man and wife. In fact, this love for each other, along with intense peer pressure, forms the powerful military concept of unit cohesion. It allows Marines to cohere and not break and run, even under the worst circumstances. Lt Col Grossman in his respected book, *On Killing*, wrote: "Among men who are bonded together so intensely, there is a powerful process of peer pressure in which the individual cares so deeply about his comrades and what they think about him that he would rather die than let them down." "Cares so deeply" is a euphemism for love. In Marine culture, the unit becomes your family. You love Marines like brothers and sisters (and sometimes even more). And you want them to love you, to respect you. You will do anything for them, including giving your life. When the chips are down, there is only your unit and your buddies. Why the love? Part comes from shared depravation and hardship. Marines experience together, even in training, some of the toughest and most difficult things humans can endure. In fact, this shared hardship illustrates another aspect of Marine culture: the way Marines demonstrate levity and hilarity during some of the most physically arduous tasks imaginable. Why the levity, the hilarity, the playing around immediately before you experience incredible pain? Marine culture. In *Welcome to Vietnam, Macho Man*, Ernest Spencer talked about how joking around before battle was part of Marine Corps' tradition:

> We are grinning and joking…That's another Marine tradition, joking around when you're going into battle. I do it because I don't want to lay any of my shit off on others. If I can make others laugh, I will. Who wants a sourpuss when you're walking in and you know someone's not going to walk out? I get real serious as soon as I cross that bridge, though. Yep, my asshole jerks about two notches tighter. I am a salty old veteran, but my asshole is still young.

At TBS, all lieutenants had to run the Endurance Course, a true ball-buster of a run, with Marines being timed on a 6.4 mile run over steep hills and difficult obstacles, wearing a helmet, deuce-gear, ALICE pack, and combat boots, and all while carrying an M-16A2. Well, right before the run, which everyone had previously run without being timed, Marines were "smoking and joking" while standing in line. They were laughing, telling jokes, pushing each other around. Kind of like bear cubs playing around, rolling in the grass, swiping each other, having a good time. And they all knew that physical torture to their bodies was imminent.

This frivolity also applies to field exercises where tactics are taught. The Marine Corps tries to approximate the stress and pressure of battle. Consequently, when it does this, you see the same comical behaviors rise to the surface. Personally, I have never laughed so hard in my entire life as I did while in the Marine Corps schools. Marines are, by and large, the funniest, most down-to-earth people I've ever met. And many times they do extremely stupid things, which, given Marine culture, also make them extremely funny things.

In Marine culture, it doesn't take long to know the true nature of everyone, just like a family. Because you're around them all the time, you see the best and worst sides of Marines. You see how they behave during adversity, how they treat others, how selfish they become, and the like. As a result, like a family, Marine culture imbues a high level of candor in its Marines. Marines are blunt. And they are perceptive. You can't fool them. The bottom line is that no Marine can hide himself from other Marines. He can't hide his exterior, he can't hide his interior. Instructors continually say: "Devildogs, you can't hide here. Not from us, not from your fellow Marines. Just can't do it."

Marines have an intense personal desire to be with their unit (usually meaning their platoon and specifically with their squad). Marines don't want to be alone or miss anything. They want to be with the unit and do everything with it. All the time. If Marines are left out of a platoon activity, regardless of the reason, they feel like Outsiders, and they have an overwhelming need to get back to the unit, immediately. This also applies to things that make them different from the unit—things like forgetting a piece of equipment when no one else did. Marines don't want to be different.

19

Peer Pressure

One pervasive aspect of Marine culture is peer pressure. Normally, in the civilian world, the concept of peer pressure, commonly introduced to people in grade school, is overwhelmingly construed in a negative sense. But in Marine culture, it has significant positive connotations. For example, it is the one force that contributes mightily to Marines cohering on the battlefield and not cutting and running in the face of the enemy. The plain fact is: Many if not most Marines on the battlefield are more fearful of being called a coward and subsequently being ostracized by their brethren than they are of being wounded or killed by the enemy. Admittedly, peer pressure is not the only thing that leads to Marines cohering under extreme circumstances; there also is shared deprivation and love for one another, military training, military discipline, and the Uniform Code of Military Justice, which prohibits cowardly conduct. But peer pressure is still a major contributor. On the other hand, peer pressure has negative connotations in Marine culture. It creates a sense of constant approval-seeking behavior. Not only that, it also produces bullies who use peer pressure as a tool to not only make another Marine conform to custom and tradition, but also to exercise power and make that Marine conform to the wishes of the bully.

During TBS, a squad of Marine lieutenants was being transported from one training area to another in a "cattle car," a euphemism for a troop-transport truck. In the cattle car, in route to the training area, the conversation among the Marines turned to masturbation, a common occurrence in Marine culture. (Marines love to talk and joke about masturbation. They call it, among other things, "whacking," "punching your clown," and "beatin' off.") On this particular day in the cattle car, Marines were engaged in a spirited discussion—if not a dissection—of the fine art of male masturbation. One lone lieutenant, however, did not appreciate the discussion and, after a short while, got irritated and decided to voice his disapproval. Above the din of masturbatory conversation, he yelled for everyone to cut out the masturbatory talk. He asked why everyone feels the need to always talk about masturbation. Wrong move. The squad, the pumped up crowd, turned on the unwise lieutenant and basically had a field day with him. They said, "Aw, are we hurting your feelings?...Maybe we're hitting too close to home, hungh?...Does mommy disapprove?...Are you a closet masturbator?" And on and on. From that moment on, the lieutenant was privately and publicly spoken about only in a pejorative sense. Later on, most of the Marines in the cattle car that day would privately say that the Marine exercised bad judgment to speak out like that. For me, the lesson was clear: don't do what Marines don't want you to do—or else. Peer pressure is powerful and it can make or break your reputation and standing among Marines.

20

Physical Appearance

Physical appearance goes a long way in Marine culture. You see, Marines prefer a certain look for themselves: Square jawline, big arms, bull neck, V-back with a bulging trapezius, washboard abs, deep command voice. The whole bit. All these things give a Marine, The Look. Of physical strength. Of toughness. Of credibility. But Marines also like athleticism, which is slightly different than The Look. Marines who are athletically inclined—fast, strong, agile, good PTers—become known as PT Studs. And these PT Studs, if they also have The Look, are anointed with credibility and respect in Marine culture, even before their peers and juniors get to know them. Imagine NFL Hall of Famer Howie Long coming before a platoon of Marines to speak to them as their commanding officer. Because Howie is big, muscular, athletic—not to mention famously handsome—and has The Look in bountiful quantities, Marines would anoint him immediately, even before he spoke. Unfair? Perhaps. Because maybe, after awhile, Howie wouldn't measure up and be a "good Marine." It could also be unfair in the reverse, too. Marines sometimes don't get anointed immediately, because they don't have The Look or aren't athletically inclined, even though they might be exceedingly competent leaders. However, that doesn't mean these Marines can't (or couldn't) get the anointment; they just have a steeper hill to climb. But when viewed in the aggregate, perhaps all this is fair. After all, many times the anointment is accurate, since that particular Marine may have The Look, is athletically inclined, and is a good Marine. Furthermore, the anointment is just a rebuttable presumption. In other words, the anointment disappears if the Marine screws the pooch by not being a good Marine, not looking after his Marines, not being competent, and so on. The presumption bursts and the aura of credibility and respect falls away. Just as if he had looked like Woody Allen.

Marine culture instills in Marines that they are marketing reps—walking billboards—for the Marine Corps. This is because when civilians see Marines they often see, like it or not, the Marine Corps. And they draw inferences from a Marine's appearance and conduct, which are then imputed generally to the Marine Corps. Human nature. And make no mistake: the Marine Corps cares about its image. They also care about professionalism and how it equates to dress and conduct. Consequently, the Marine Corps places many restrictions on the appearance and conduct of its Marines on- and off-base. One area is clothes. All Marines dress alike. On-duty, Marines wear uniforms to be uniform. But Marines dress alike off-duty, too. The dressing style off-duty is called "casual." Not "civilian casual," where you wear baggy jeans, holes in the knees, and T-shirts that say "Roll Your Own" on the front, but Marine casual. Which means a respectable appearance—one that isn't unbecoming of the Marine Corps. Whatever that means. Business casual.

There are more restrictions than just clothes. Some restrictions are on the face. Unlike the Navy, no beards are allowed in Marine culture. However, mustaches are permitted, provided they are severely cropped (per regulations), with no hair hanging over the edge of the upper lip or on the corners of the mouth. All this usually leaves, when the clipping is done, a Hitler-type mustache. Because of this unsightly appearance, most Marines—enlisted and officers—opt out. Officers have another reason to opt out: tradition. Mustaches are disfavored for officers. Seeing officers with mustaches is a rare occurrence. And when you do see them, they will usually be "prior service" Marines—known as "Mustangs"—meaning, the officers first went through the enlisted ranks before becoming officers.

And then there is the matter of shaving. Like many things in the Marine Corps, shaving is a daily necessity. For all Marines. No exceptions. And this applies to weekends, too, if Marines intend to come aboard a Marine Base. In addition to having to shave, Marines have to watch their head hair. A typical haircut for a Marine is called a "high and tight," where the sides of the head are shaved and the hair on top is no longer than three inches. Infantry grunts, Marines in training, and newly minted Marines typically wear this hairstyle. Others, such as those in the air wing, where personal appearance standards are looser—"Swing with the Wing!" is their motto—let their hair grow longer. But don't think "longer" means "civilian" long. "Long" in Marine culture means seeing hair on the sides of heads instead of skin. Curiously, especially in the enlisted ranks, Marines who are "salty"—those who have been around a long time—like to grow their hair a little longer.

All these restrictions on personal appearance, as one can imagine, produce nice, neat, clean Marines. Except for the issue of tattoos. Which are, curiously, allowed. Maybe the Marine Corps likes the macho image (or they like it as a form of "posturing," harmlessly intimidating others). Even though allowed, there are some restrictions on tattoos, mainly traditional. Like it does with mustaches, Marine culture and tradition frowns upon officers having tattoos. As a result, officers don't typically have tattoos, unless they have ones pre-dating the Marine Corps experience. Of course, there still are some officers who have tattoos. But there aren't many of them (they are usually infantry officers and the tattoos are placed in inconspicuous spots, like the outside of a calf, near the ankle, on the inside of the bicep—not on the shoulders or forearms or back). For enlisted Marines, it's totally different. Tattoos are permitted. Consequently, many enlisted Marines have tattoos on their shoulders, back, forearms, chest, and legs. Tattoos (and mustaches) illuminate the cultural differences between officers and enlisted Marines.

21

MacGyver Instinct

During the *MacGyver* television series, its main character frequently created and fixed complex products and machines by using ordinary household items, such as duct tape, credit cards, lightbulbs, and even fan blades. Well, this same characteristic is internalized by Marines. And there are good reasons for it.

Through the years, despite its battle record and reputation, the Marine Corps has had to fight for its institutional life more times than it cares to remember. So to preserve its life, it has become known as a frugal, low-cost, high-bang-for-the-buck force. Robert Kaplan, in his book, *Imperial Grunts*, said that "The Marines ha[ve] always been the poorest and scrappiest of the armed services." Says Lt. Gen. Victor Krulak in *First to Fight*:

> [T]he United States Congress sees the Marines as a frugal and altogether reliable investment, dedicated, like nineteenth-century British general Sir John Moore, to "fighting on the cheap."

As part of its nature, the Marine Corps creates and fosters a mindset of penny-pinching creativity, improvisation and innovation. It inculcates a "can do" attitude, forcing Marines to make do and overcome any and all obstacles with the men, weapons, and equipment on hand—even if they're antiquated, underpowered, and useless.

One sees this in the Marine inventory. Frequently, Marines are the last service to get the latest and greatest equipment: For many years, Marines only had M-60 tanks, while the Army was tooling around in the new M1 and M1A1 Abrams tanks. For years, Marines used the old "782 gear" (web belt, ammo pouch, canteen, suspenders) as their individual load-bearing equipment, while the Army was wearing new and fancy load-bearing equipment; Marines used old and antiquated shelter-halves instead of new Army tents; Marines slept in old, heavy sleeping bags, while the Army slept in new, light, down sleeping bags—the list goes on and on. But that's not to say Marines *always* lag behind the other services in getting new weapons and equipment, because they don't. For example, Marines were the first to test and adopt the M-16A2; and, more recently, they were the first to adopt the digitized camouflage utilities worn by many Marines in Operation Iraqi Freedom II. But these are aberrational examples. Fact is, overall, the Marines, as Tom Clancy said, "take the pieces that are given to them, arrange them in unique and innovative ways…and throw in their own distinctive magic." It has given them the reputation as being penny pinchers, innovators, and "MacGyvers."

22

Promptness

Being late in Marine culture is not good. The military term for this is UA or Unauthorized Absence (the term "UA" is now used instead of "AWOL" or Absent WithOut Leave). And it applies not only to deserters and Marines who are gone for days, months, and years, but also to those who are late by mere seconds. It doesn't matter. To the Marine Corps, you are UA. And there are no acceptable excuses, either. You should have planned, like a good Marine, for the unforeseen circumstances of a traffic jam, bad weather, and the thing that supposedly made you late. Being late rankles other Marines, too, not just a Marine's seniors, and they'll call you on it. A tardy Marine's peers will pointedly say, "You're UA!" in front of others at a formation (others may be silent, but they are thinking the same thing, drawing adverse inferences about that Marine's character). If late for a meeting, a senior Marine will highlight your tardiness in front of others, or tell you in private afterward. But be rest assured: you will hear about it. Why? Again it gets down to details, and the enormity of the stakes in the military. People could die because of your tardiness, because of your inattention to detail, because of your selfishness. Being one minute late could cause the death of a unit waiting to be reinforced in enemy territory. Consequently, Marines err on the side of alacrity and promptness, not sluggishness and tardiness. Marine culture embraces promptness, rapidity of movement, and perfectionism.

Promptness

To put it clearly, under the Marine Corps' twisted logic, if a Marine is on-time for an event, he's late. (Some Marines will even say: "If you're 10 minutes early, you're 5 minutes late.") The rule to many Marines is this: Being on-time means being at least 15 minutes early.

23

Courtesy

The amount of courtesy demanded from Marines is considerable—much more than a civilian can possibly imagine, and even more than the other military services muster on a daily basis. On a sliding scale, there is courtesy, military courtesy, and Marine courtesy. And there's a world of difference between all three. In Marine culture, prodigous amounts of "Sirs" tumble from open mouths (by the way, only officers are called "Sir"; enlisted members are called by their specific rank—Lance Corporal, Corporal, Sergeant, Staff Sergeant, and so on.) Every time you talk to an officer, you use "sir," either at the beginning or end of a question or statement, or at the beginning or end of an answer. And the "Sirs" don't just tumble on-duty, they tumble off-duty, as well. There are no boundaries when it comes to Marine courtesy. (The Uniform Code of Military Justice, known as the UCMJ, codifies and creates a legal duty to be respectful to senior ranking officers, and it doesn't differentiate between being on- and off-duty, or being in or out of uniform.) And "off-duty" can be anywhere, in a bar, restaurant, grocery store, laundry-mat. Doesn't matter. Marines are compelled to be courteous and use "Sir" (unless the person was pulled aside and told not to use "Sir"—which, of course, doesn't happen much).

Being courteous and respectful does not, by itself, make a good Marine. But it's an indicator of a good Marine (it's a high honor in the Marine Corps to be branded with the appellation, "good Marine"). In the view of many, good Marines don't think courtesy and respect negate toughness and courage. They know these values are not mutually exclusive. And they know these values are part of what creates a "good Marine." When you look around the Marine Corps, it soon becomes obvious that the good, tough, hard-working Marines are generally the ones who exhibit the most courtesy and respect. It is the reverse for bad Marines. There are exceptions, of course. Discourteous, impolite, insolent, and derisive Marines can be great warriors in battle, and thus good Marines. Respect doesn't negate courage any more than disrespect negates courage. We are talking about generalizations here. And in generalizing, good Marines are typically respectful and not disrespectful.

24

Honesty

In Marine culture, a huge emphasis is placed on honesty. In fact, it's an omnipresent emphasis. This results in a Marine's word being trusted. End of story. No checking with another person. No corroboration needed. When a Marine Officer says something, other Marines act on it, typically unquestioningly. Why the trust? Some say a lot has to do with the UCMJ, which makes it a crime to utter a false official statement, either orally or in writing. But another thing was peer pressure and the threat of ostracism. If you are caught in a lie, even a small one, it spreads through the small Marine community like a bad cold. And if a Marine fears anything, it is ostracism. In fact, the last thing a Marine wants is to become a social pariah.

Underlying the spoken and written word was trust. Trust by other Marines. Trust that the words are true and accurate. Trust in the speaker. Because of this trust, Marines usually rely on the words of other Marines without hesitation and without corroboration.

As the speaker, one feels a great responsibility not to betray this trust. Even in small ways. Marines don't tend to embellish or embroider stories. They stick to the facts. Straight and direct and accurate. It is marvelously refreshing.

25

Rank

In the Marine Corps, rank is very important. Indeed, it is the ultimate status symbol. In the civilian world, status typically equates to money, with civilians aspiring to accumulate it. In Marine culture, money is not nearly as important. After all, people don't go in the military to make lots of money. They do it for other reasons: patriotism, a test of manhood, power, and the like. Rank replaces money in Marine culture. An incredible amount of respect comes with rank, much more than in the civilian world. And it becomes greater as the rank increases (and it also becomes greater with one's years of service, which permits one to be called "salty").

Like most things in Marine culture, achieving rank is extraordinarily competitive. In fact, advancing in rank is one of the most difficult enterprises in the Marine Corps, especially for officers. Of the military services, the Marine Corps has the lowest officer-to-enlisted ratio (approximately 1 officer to 8.8 enlisted). What is more, it has an up-or-out policy, which involuntarily separates Marines "passed over" twice for promotion (in practical terms, getting passed over once almost guarantees one of getting passed over twice and thus kicked out of the service). Nevertheless, climbing the ziggurat of rank is competitive, and it shows itself in small ways. To climb higher, most Marines seek opportunities to command. Even small ones. For instance, when an Officer-In-Charge (OIC) goes on leave or assignment, or is out of the area, the remaining officers—who are many times similarly ranked—jockey for position and the opportunity to command. Each is trying to be the lead dog. Why? Getting "command time" helps a Marine's comparative ranking on his fitness report and it helps him gain an edge with promotion boards. Who gets picked for the command opportunities? Usually the Marine who has the most years of service. And the date that controls is the Marine's active duty date, i.e., the date he came onto active duty in the Marine Corps. The earliest date wins. An earlier date, even by one day, means a Marine is "senior" to another, and therefore he obtains precedence in obtaining command (the terms "senior" and "junior" also apply to rank, not just years of service). And you'll actually hear and see this: two or three similarly-ranked officers, standing in the vacuum of their OIC's absence, comparing dates, getting wrapped around the axle about it, sorting out who's the lead dog based on years of service. Not competence. Not merit. Years of service.

Here's another example, albeit a small one, showing the emphasis Marines place on rank. Many Marines will place their rank on their car or truck (known as a "POV," Privately-Owned Vehicle). Enlisted Marines are seemingly more prone to do this, although some officers also do this. They go to the Base Exchange, purchase a colorful gold and crimson sticker in the form of rank chevrons (or officer insignia), and slap it on the window of their vehicle (sometimes on the front windshield, sometimes on the rear window, sometimes on the bumper). The funny thing is, Marines don't think this is unusual. It's merely part of their culture. Rank is status, and it's OK to flaunt it—even if it's an ostentatious display. A great many POVs in the Marine

Corps have these stickers. In the civilian world, this would probably be analogous to placing a sticker that says "Chief Executive Officer" on one's car bumper. Doesn't typically happen. True, sometimes civilians get vanity license plates that have "CEO" (or some other stuffy title) on them, which is somewhat analogous to the chevron bumper stickers in the Marines, except for one small detail: the sheer per capita number of rank stickers in Marine culture.

At Marine officer training schools, a student Marine lieutenant has no authority, despite his rank, and he barely rates respect. Even though TBS students are commissioned lieutenants, they are still students. They joke that TBS is the only place in the Marine Corps where a lieutenant receives orders from Lance Corporals. Those Lance Corporals, known derisively to student lieutenants as Lance Colonels, only have to precede or follow the order with "Sir" or "Gentlemen" to get away with saying most anything and, more importantly, ordering lieutenants around. Example: "Gentlemen, those cigarette butts on the ground need to be placed in the trash can," or "That weapon looks terrible, Sir. It's far from clean; you'll have to scrub it some more before I'll permit it back into the armory, Sir." On the other hand, Marine lieutenants are expected to be respectful to anyone who is their senior, which seems like everyone on a base like Quantico, where OCS and TBS are located, since it is satiated with high-ranking officers, probably more than any other Marine Corps Base.

A word about "class discrimination," which is another startling attribute of Marine culture. Before specific examples are given, one must first remember that the Marine Corps discriminates frequently, and most of it is lawful and justified. Not everyone can or should be a Marine. Fact is, there are legitimate base-level mental and physical prerequisites one must meet before becoming a Marine. And if you can't meet them, too bad, so sad. This is understandable. This is expected. But the Marine Corps also lawfully discriminates based on the status or class of Marines themselves, i.e., between officers and enlisted personnel. In fact, there is a huge demarcation between officers and enlisted personnel, in spirit, in regulations, and in reality. And sometimes the demarcation results in restrictions that may seem reminiscent of the old Jim Crow laws of the South, the ones that restricted the use of public facilities to "Whites" and "Coloreds." In the Marine Corps, it is not uncommon to see signs on heads (restrooms) saying the following: "Officers and Senior Enlisted Personnel Only" and "Enlisted Only." It is also not uncommon on Marine Bases to have two separate clubs for Marines: one for officers, called the Officers' Club, or "O Club" for short; and one for enlisted folks, called the Enlisted Club, or "E Club." Officers and enlisted members were not allowed to cross-pollinate in these clubs, except when required by special circumstances, i.e., when a Marine Corps Ball was taking place or when officers and enlisted were needed to perform ceremonial functions.

26

Institutional Hallmarks

The Marine Corps is an institution that has a song (the Marines' Hymn), a motto (*Semper Fidelis*, i.e., "Always Faithful"), a birthday (November 10, 1775), and an emblem or device (the Eagle, Globe, and Anchor).

The Marines' Hymn contains words that are so clear, so accurate, and so motivating, that whenever it is played, Marines get to their feet and stand at attention until it ends. All of which produces goose bumps, almost without fail. (There's a Marine myth that if you don't get goose bumps when the Marines' Hymn is played, you are not a real Marine.) The Marines' Hymn is typically played at Birthday Balls and changes of command. At the end of the Hymn, Marines will bark and ooh-rah at high decibels.

The author of the Hymn is unknown. The Hymn comprises three stanzas in its full form, although the last two stanzas are largely unnoticed:

> From the Halls of Montezuma
> To the shores of Tripoli,
> We fight our country's battles
> In the air, on land, and sea.
> First to fight for right and freedom,
> And to keep our honor clean,
> We are proud to claim the title
> Of United States Marine.

On Marine Culture

Our flag's unfurl'd to every breeze
From dawn to setting sun;
We have fought in every clime and place
Where we could take a gun.
In the snow of far-off northern lands
And in sunny tropic scenes,
You will find us always on the job—
The United States Marines.

Here's health to you and to our Corps
Which we are proud to serve;
In many a strife we've fought for life
And never lost our nerve.
If the Army and Navy
Ever look on Heaven's scenes,
They will find the streets are guarded
By United States Marines.

The Marine Corps' motto is *Semper Fidelis*, meaning "Always faithful." Although this motto gets a lot of play in the American media, and is well known by civilians, it is rarely spoken or written in Marine culture—at least not until a Marine leaves the Marine Corps. And then, once on the outside, it is OK and even expected to sign off on letters and e-mails with "S/F" or "*Semper Fi*." While on the inside, Marines are too macho to go around saying *Semper Fi* to other Marines, which is tantamount to saying, "I love you, brother and I'll always be faithful to you. Always! I promise!" This is not to say that Marines don't get maudlin about their brothers and brotherhood, because they surely do. They just don't use *Semper Fi* to show it. But Marines do contort

Semper Fi into other phrases, like *Semper I*, to highlight another Marine's selfishness or actions that are antithetical to the Marine team, or *Semper Gumby*, to illustrate how Marines must be flexible and adapt to any situation at hand.

The Marine Corps celebrates its founding at Tun Tavern in Philadelphia on November 10, 1775 by having an annual birthday party. A birthday party? Yes. Although Marines join their sister services in honoring Armed Forces Day every May, Marines take their institution's birthday far more seriously. So every year, on November 10, Marines across the globe pull on their dress uniforms and engage in celebration and comradeship. They walk around and shake each other's hands and say, "Happy Birthday, Marine." It makes every Marine feel special, like he is part of an elite team. Marines also get invited to Birthday Balls put on by their Command, which usually occur on or around November 10, and involve wearing Dress Blues, honoring the youngest and oldest members of the Command, cutting the Marine Corps' birthday cake, imbibing on much alcohol, and smoking cigars. Most Marines bring dates to the Balls, known affectionately as "ball dates." Finding a ball date, unless a Marine is married or has a local girlfriend, is numero uno priority around October (and the first week of November) every year.

The Marine emblem—the EGA—is awarded to all Marines upon graduation of recruit and officer training. One sees it emblazoned in black on Marine covers, helmets, and also on the front left pocket of cammie blouses. One also sees it, in three-dimensional form, affixed to Marine dress uniforms on the lapels and covers. The eagle is supposed to signify America and freedom; the globe is supposed to signify the Marine Corps' worldwide capability, having 7 horizontal lines across it, representing the world's 7 continents; and the anchor is supposed to signify the Marine Corps' maritime nature, coupled as it is with the U.S. Navy.

27

Marine Monikers

Throughout their history Marines have acquired numerous monikers beyond the appellation "Marine." For example, they have been, and are, variously referred to as Leathernecks, Devildogs, and Jarheads. Each of these monikers has its own genesis or etymology.

The term "Leatherneck," according to *The Marine Officer's Guide*, "goes back to the leather stock, or neckpiece, that was part of the marine uniform from 1775 to 1875." Although some say the tall leather stock was meant to protect a Marine's neck in a sword fight, perhaps a common experience in the old days, the *Guide* says that "the truth seems to be that it was intended to ensure that Marines kept their heads erect…a laudable aim in any military organization, any time." This tall stock has been carried down by tradition and is visible even today in the Marine Dress Blue uniform.

The term "Devildog" was conferred in 1918 during World War I by the German Army after it fought the Marines at the battle of Belleau Wood in France. According to the Germans, the Marines fought so tenaciously, so remorselessly, in defeating them that they appeared to fight like *Teufelhunden*, or dogs possessed by the devil. As an aside, the French were so euphoric that the Marines saved the day after their own troops retreated, they re-named Belleau Wood "The Wood of the Marine Brigade." It is common lore in the Marine Corps that if a Marine knocks on the door of a house in that area of France, he will be treated like Royalty, or, even better, to dinner.

The term "Jarhead" is of course less flattering than the first two. Depending upon who you talk with, "Jarhead" can mean the way a Marine's head looks after he gets a high-and-tight haircut, or it could mean that a Marine's head is devoid of a brain, like an empty jar.

28

Marine Speak

Marines generally speak more directly and forcefully than people from other services, and surely more forcefully than civilians (the Marine Corps even has a category on their Fitness Reports called "Force"). Marines get right to the issue. And while getting to the issue, they do more than just talk. They gesture and gesticulate for emphasis. But they don't gesture and gesticulate like a caffeinated Italian, pinching fingers and thumbs together, throwing around their hands and forearms, causing people to duck. No, that wouldn't be right. Marines are too squared-away, too controlled and precise, too robotic. Instead, Marines move in right angles. (Truth be told, many will even march down the hallways and pivot on the ball of one foot when turning a corner, just like a good marching Marine. Many Marines do this instinctively, almost like they are having fun.) Instead of flailing around, looking undisciplined, Marines karate chop the air while talking. Karate chop? Yes, they will cleanly karate chop the air in front of them with one hand, in effect saying, "That's the way it is!" or "It's over there, in that direction!" And for emphasis, they'll karate chop their palms. As one might expect, this karate chop, usually demonstrated with the dominant hand, has the thumb flush against the index finger, with the rest of the fingers evenly pressed together; the other hand, the non-dominant hand, has its palm upturned and flat, creating a rigid landing zone for the forceful chop. Of course, this landing zone has all fingers (and thumb) nicely pressed together, just like the karate chop hand. So the karate chop is all controlled force. Nice and neat. No excess parts. All right angles. No one ducking for cover. Now at other times, Marines will talk with their hands on their hips using a manly command voice and a steady gaze. Like John Wayne. Then, for emphasis, they will bust out a karate chop straight from the hip, like a quick-draw.

Some say the karate chop and precision movements are inspired by the time spent drilling in formations on the parade deck. This could be true. But what is also true is that not all Marines karate chop and have precise movements. This could be because someone doesn't want to be like everyone else. Or because of an anti-authoritarian streak. Nevertheless, the Marines are filled with many types, the karate choppers, the precision movers, and the less precise non-karate choppers. But the former prevails in numbers over the latter.

Marine culture produces a lexicon of words and phrases, which often times imperceptibly creep into the Marines' vernacular. When they do, they are passed down from one generation of Marines to another. Consider the following words and phrases:

"**Aye, aye, Sir!**" (received, understand, and will execute the order, Sir!)

"**Bark, bark**" (a play on "Ooh-Rahh," the traditional version of the Marine bark)"

"Wear your **BCGs**" (Birth Control Goggles or military-issued glasses/spectacles)

Marine Speak

"I need my **black Cadillacs**" (combat boots)

"We're supposed to wear **boots and utes**" (combat boots and camouflaged utilities)

"He was **born hard**" (he's a tough, no-nonsense, uncompromising Marine)

"He's in the **Brig** (he's in jail; "brig" means a type of sailing ship, which was used historically as a jail when moored in a port or bay with sails furled)

"**Buddy is half a word**" (as in "Buddyfucker")

"Watch out for the **bulkhead**" (watch out for the wall)

"**Check or hold**" (understand or not)

"You'll need a **chit** for that" (an authorization or a receipt)

"Let's get some **chow**" (go eat)

"How about going to the **commissary**? (base grocery store)

"Get on the **deck**" (get down on the floor)

"Got my **dick in the dirt**" (working hard, usually out in the field, doing what Marines do best)

"He's a **doggie** or **dogface**" (a soldier in the Army)

"He's a **fatbody**" (not as slim as he should be)

"Thursday is **field day** in the Marine Corps" (a time set aside for the cleanup or policing of a general or specific area)

"Are you **gaffing me off**? (Are you ignoring me?)

On Marine Culture

"**Get some!**" (tough it out and execute the mission)

"I need my **go-fasters**" (running shoes)

"What's the **gouge**?" (inside information)

"I'm gonna get a **grape scrape**" (a haircut, usually meaning a high-and-tight haircut)

"**Grow a pair**" (same thing)

"He's **gung-ho**" (he's a hard charger and has *esprit de corps*)

"It's **gut-check** time" (it's time to see who can tough it out)

"He's a **hardcharger**" (a good Marine)

"Close the **hatch**" (close the door)

"I have to use the **head**" (go to the bathroom)

"That's bad **ju-ju**" (bad stuff)

"We've got **junk on a bunk** tomorrow" (we have an inspection of our clothing and equipment, displayed on our racks)

"Here's a little knowledge for your **knowledge knot**" (a little information for your brain)

"**Lock your skuzzy buddy up**" (assume the position of attention)

"Where's your **moonbeam**?" (where's your flashlight)

"**Negative**" or "That's a negative, good buddy" (no or that's a no)

Marine Speak

"**No excuse, Sir**" (the common response from a Marine in training, when questioned about his actions)

"He's gonna get "**office hours**" or **NJP**" (he's gonna get Non-Judicial Punishment imposed by his commanding officer instead of being court-martialed)

"**Ooh-Rahh**!" (the Marine bark)

"**Outstanding**!" (a ubiquitous word in Marine culture used in response to most everything)

"**Pack your trash and saddle up**" (stow your personal gear and get on your feet).

"Do you have some **pogey bait**?" (do you have some children's candy?)

"We're gonna **pull chocks**" (move out or shove off)

"He **punches his clown**!" (masturbates)

"Wanna go to the **PX**? (the Post Exchange, a store on base)

"He's in the **rack**" (in the bunk)

"**Reach down and grab a pair**" (be a man)

"**Roger that**" (understand and agree)

"Do you have **sand in your pussy**?" (you're whining for no reason)

"What's the **scuttlebutt**?" (what's the rumor, i.e., the word on the street?)

"*Semper I*" (a play on "*Semper Fi*," implying that a Marine is being selfish)

"*Semper Gumby*" (a play on "*Semper Fi*," implying that change is continuous and that Marines always have to remain flexible—like a Gumby)

"He's a **shitbird**" (a bad Marine)

"He's a **swabbie**" (a sailor)

"He's a **weak sister**, **non-hacker**, and **slacker**" (a Marine not pulling his own weight, not living up to standards)

"He **whacks!**" (masturbates)

"She's a **WM** (a woman Marine)

"Don't get **wrapped around the axle**" (don't get all uptight and nervous)

"What's The **Word**" (the plan or orders to be followed)

"**Yarch!**" (how the word "march" sounds when calling cadence)

"**Yoo Hoo!**" (getting someone's attention)

"**Yutt Yutt!**" (I have no clue what this means)

29

History and Tradition

History is important. If we forget history, we're doomed to repeat it, as George Santayana has instructed us. Marine culture, and military culture in general, places great emphasis on history, and on tradition and rituals—otherwise known generally as pomp and ceremony. This pomp and ceremony even applies to daily duties, like raising and lowering flags, called "Colors." Every morning and evening on Marine bases there is a "Colors" ceremony, where every flag on base is raised and lowered by "Flag Details" of two to four Marines. In the Marine Corps, you just can't send out a PFC to yank down the flag, roll it in a ball, and toss it in the corner. That doesn't quite work. So you have "Colors." The "Colors" ceremony lasts maybe 10 minutes. And five minutes before "Colors" is commenced, you'll get a five-minute warning: a short burst of music played from speakers around the base. And then during Colors, the music blares and plays till the end of the ceremony. If outside at the time, regardless of rank, and regardless of exact location, every Marine has to stop, face the nearest flag (even if you can't see the thing), come to attention, and then salute—and you must hold the salute for the duration of "Colors"—or until the music stops (to avoid having to do this, Marines will linger in buildings until "Colors" ends). Cars will even be stopped by Road Guards near the flag, their occupants compelled to exit and salute the flag. The flag, more than everything else, with the possible exception of the EGA,

Congressional Medal of Honor (and similar medals for valor), is the most revered symbol in Marine culture.

In Marine culture there is an unending schedule of ceremonies: for changes of command, retirements and notable anniversaries. These ceremonies also include parades and therefore require "parade practice" by many Marines, who unfortunately get tagged with such duty. Many Marines hate Parade practice. It usually requires hours of daily commitment, over and above regular duties, and frequently includes the drawing of gear from the supply shack or weapons from the armory before practice. And then it is marching, yelling commands, and drilling for at least two hours—and many times under a terribly hot sun. All for a parade that may last half an hour.

When I first reported to Camp Pendleton in August 1992, I was tagged as "Platoon Commander" for the 50th Anniversary parade at Camp Pendleton. It was two weeks of practicing two hours a day, on top of a busy legal schedule. I was a brand new lieutenant, just out of TBS and Naval Justice School, standing before a platoon of 38 enlisted Marines. These Marines, who stared hard at me, wore tattoos and big black G-shock watches (also part of the culture). I remember thinking that this was one scary bunch of dudes. After two weeks of marching, yelling commands and drilling, D-Day came. The 50th anniversary parade was over in an hour. With not five minutes of marching. Twenty hours of drill for five minutes of marching. This is the Marine Corps' way.

History, of course, has more uses than just knowledge and pomp and ceremony. If used properly, it is a great motivator. Of course the Marine Corps realizes this, and provides historical reminders. Statues, memorials, and monuments are conspicuously placed around Marine bases. All about personal sacrifice. All about the rewards of selflessness. All about motivation. There are also simple things like street signs bearing the names of famous Marine battles—Iwo Jima, Guadacanal, Tarawa—and famous Marines—Puller, Butler, and Vandegrift—and so on.

As part of my PT routine in the Marines, I would run along these roads and trails. I would notice the street sign, think about Marines overcoming inestimable odds in that particular battle reflected on the street sign, and would immediately get goose bumps. I would feel the endorphins releasing, giving me bursts of strength, making me run faster, as if I heard a great song come through my radio's earbuds. I had a rush of pride, knowing I passed muster: I was a Marine just like the Marines of yesteryear who overcame incredible odds and accomplished so much. Tom Wolfe coined the phrase "The Right Stuff." At times like these, I felt I had the "Marine Stuff." And all because I looked at a street sign.

30

You Can Take the Man Out of the Corps, But You Can't Take the Corps Out of the Man

I'd like to close my discussion on Marine culture as I started it: by recognizing the deep and often permanent effects that Marine culture has on the behavior and conduct of its Marines. Many if not most Marines who leave the Marine Corps think they have left it behind. They are wrong—at least in the psychological sense. While they may leave the Marine Corps *physically*, they do not leave it *psychologically*. In fact, Marine culture's formless and limitless branding ways follow and impact them wherever they go—even until they die.

I was profoundly affected by Marine culture. I didn't fully realize it then, however. Only after I got out did I come to fully realize and appreciate it. I needed the passage of time. And cool reflection. I needed to decompress and to compare my Marine Corps' experience to something else, like the civilian world. It lasted a year, maybe two. In sum, I realized that Marine culture had changed me in a number of significant and profound ways. What follows is my discussion of how I was changed, for better or worse. On a personal note, it was often difficult to specifically identify a change in my personality, behavior, or conduct that could exclusively be attributed to Marine culture. After all, a person's basic nature and personality are formed well before he enters the

Marine Corps. And the Marine Corps, and specifically its culture, refines, strengthens, weakens, and changes one's pre-existing qualities and traits. But sometimes it creates brand new qualities and traits; at other times it eliminates pre-existing qualities and traits. All this cannot be ignored. The difficult part is in determining whether the change is because of the Marine Corps, your advancing age, or your naturally changing personality and character. Here's my shot at it.

First of all, Marine culture gave me a relational experience to put my personal problems into perspective and proper focus. When I encounter problems in civilian life, they almost always pale in comparison to the trials and tribulations of Marine officer training and being a Marine officer. Having this perspective, this relational experience, helps me immensely when the going gets tough. I compare and contrast my experiences frequently. And when the comparison is complete, I am typically energized with a sense of unwavering confidence in my own abilities to overcome the obstacles.

Marine culture made my style of communication, both orally and in writing, more direct and no nonsense. In terms of public speaking and leadership, I always remember what I was taught by the Sergeant Major of TBS during Drill exercises on the parade deck: There is a direct relationship between the forcefulness of your speech and the quick and exhuberant obedience of your audience. And it's true. I saw it time and time again on the parade deck. Speak confidently and forcefully, and people react quickly. Speak meekly and hesitatingly, and people react, if at all, slowly. People play off your words, your tone, and your force. And it just doesn't apply in the Marine Corps; it applies in civilian life, too.

Marine culture made me appreciate and enjoy life, and to live life more in the present moment, laughing about matters even though they may be distasteful, onerous, and physically punishing. In the Marine Corps, especially in its training programs, your life is scheduled down to the minute. And you have so many activities and tasks to accomplish every day, every hour, every minute. It makes you focus on exactly what you are doing—not what is planned in an hour or two, or a day or two. Many times those future tasks seem like an eternity away. So Marine schedules, and Marine culture, made me live my life in the present moment, not thinking about the past consumed with guilt, or the future consumed with worry. It's a healthier way to live life, and although this attitude pre-existed the Marine Corps, I find it much easier to maintain the attitude now in the civilian world.

Marine culture instilled in me a keen appreciation for attention to detail, hard work, and fast-paced decision-making. Marines focus, focus, and focus some more on those details, to their ever-lasting credit. And that trait doesn't disappear in civilian life. Although I must say, it does have its drawbacks: the bad parts of attention to detail, becoming a "stress grenade" and "anal retentive" sometimes work to your detriment in civilian life. Your family, friends, and co-workers will raise an eyebrow at your statements and conduct and say something like: "Chill out, dude…it's not that big of a deal." Well, they haven't been in the Marine Corps. I also learned to work hard, and not become complacent. Marines know that complacency leads to mistakes and eventually to defeat. In the Marine Corps, as in life, victory typically goes to those who work the hardest, those who do not become complacent (like those who have much experience or who are smarter than their opponents). I learned from this. Finally, the Marine Corps taught me to take quick, decisive decisions. I have found that by acting quickly and decisively in the civilian world, you are frequently given an advantage over your opponents, or to achieve what you want to achieve.

Marine culture humanized me by making me understand that everyone is human, and we all suffer from and feel the effects of fear, food deprivation, water deprivation, i.e., from our base needs and reactions. It also taught me that the important thing is not to be immobilized by the fear or deprivation of base needs. Even though this may sound like a big contradiction in that the Marine Corps is about masses of men who generally are perceived to act as unfeeling, fearless automatons, Marine culture fosters an environment where showing fear, or the effects of fear, is OK. And expected. Just don't become immobilized by it. Move, move, move. Take decisions, take action. I learned from that. I frequently feel fear, and show it, in civilian life. But you know what? It doesn't stop or immobilize me. And Marine training and culture, perhaps more importantly, taught me not to be ashamed of feeling fear and showing its effects. I don't feel guilty about it. And I know I am not a coward just because I feel and show fear. It makes me a coward to be immobilized by fear.

Marine culture cultivates a high degree of trust and loyalty between its members, which facilitated making friends inside but not outside the Marine Corps. As a Marine inside the Corps, when you meet another Marine for the first time, it is easy to converse and become friends. In fact, it was frequently effortless and spontaneous. As James Webb, a former Marine officer and former Secretary of Navy, said in *American Spartans*: "When I meet a man who is a former Marine, I automatically trust him. And that trust has never failed me." Without even knowing the person, you knew you had many things in common: you're both Marines, you both went through the same training, you both shared miserable experiences, you both probably joined the Marine Corps for similar reasons, you both are probably not avaricious or money hungry, you both are honest and truthful—and so on. But when you leave the Marine Corps, things change. And many times for the worse. I personally have had a difficult time making close friends since leaving the Marine Corps and Marine culture. In the Marines, it's easy and not risky to open yourself up to another Marine and be close. Extremely close. The trust is indescribable. The comraderie is indescribable. But you don't have that on the outside— or at least not nearly to the degree that it's present on the inside. Thus, since leaving the Marine Corps, I find myself not becoming close friends with anyone. I have many friends and acquaintances, including clients and associates, but not close friends. The only close friends I currently have are those who were fellow Marines with me, or those who were my friends before I entered the Marine Corps. But in all seriousness, is this caused by my advancing age? Would this have happened even without my experience in the Marine Corps? It is an unknowable.

Marine culture instills physical training as part of your being, your psyche. Before entering the Marine Corps, I stayed in pretty good shape, mainly by lifting weights and running—but the running was periodic, not systematic, and certainly not daily. The Marine Corps changed that. Now it's the reverse. I now run systematically, usually 3-4 times a week, 3-4 miles on each day, and lift weights periodically (but with pushups before or after every run). I learned in the Marine Corps that staying in shape is a virtue: it keeps you healthy, happy, more alert, more energetic, and it makes you look younger, sometimes much younger, than similarly aged civilians. On this last point, one need only look at high-ranking Marine officers, those ranked Major, Lt. Colonel, Colonel, and General. The vast majority of these officers are frequently in their 40s or 50s, and when compared with similarly-aged civilians or officers from other services, there is just no comparison. The Marines, almost without exception, will be extraordinarily fit and will look extraordinarily young for their age—much more so, on average, than civilians and service members from other branches. In fact, most will look at least a decade younger than their actually age. Now of course I'm speaking generally, and there are exceptions. But there is enough disparity to make a general rule. Now, another thing I learned from Marine culture is not to give up. This can be shown in PT events and running. You learn, and I have learned, that you don't stop or give up, no matter how much you're hurting. There are two reasons to stop or give up in the Marine Corps: if you're dead or paralyzed. People who dropped out of runs in the Marine Corps were known as "run drops," "non-hackers," and "quitters"—sobriquets tantamount to calling someone a coward. At TBS, my

On Marine Culture

Platoon Commander, 1st Lieutenant J.M. Bright, a Bronze Star (with Combat "V") winner in Operation Desert Storm, told us this before a long 18-mile forced march: "Devildogs, you can be a hump-drop if any of the following occur: (1) You're walking on stumps; (2) Your intestines are coming out your anus; or (3) You're wearing a sign around your neck that says, 'I AM WEAK.' Carry on. Bark, bark."

I learned the "MacGyver instinct." In civilian life, my weapons of choice are twisty ties, duct tape, and Super Glue. When something breaks, or needs fixing, my first thought is to "MacGyver" it, by deploying my weapons of choice. And this is motivated not by frugality, but by having the MacGyver instinct drilled into me by the Marine Corps.

Marine culture made me live life with more integrity and honesty by always (or at least always striving) to do the right thing and to tell the truth (and not shade or color the truth, and not to tell just part of the truth but the whole truth). This is one of the most powerful characteristics that Marine culture instills in its members. And I was a recipient.

Semper Fi!

Bark, Bark!

www.ingramcontent.com/pod-product-compliance
Lightning Source LLC
Chambersburg PA
CBHW071707210326
41597CB00017B/2377